BUDDHIST BIOLOGY

Buddhist Biology

ANCIENT EASTERN WISDOM MEETS MODERN
WESTERN SCIENCE

David P. Barash

OXFORD
UNIVERSITY PRESS

OXFORD
UNIVERSITY PRESS

Oxford University Press is a department of the University of Oxford.
It furthers the University's objective of excellence in research, scholarship,
and education by publishing worldwide.

Oxford   New York
Auckland   Cape Town   Dar es Salaam   Hong Kong   Karachi
Kuala Lumpur   Madrid   Melbourne   Mexico City   Nairobi
New Delhi   Shanghai   Taipei   Toronto

With offices in
Argentina   Austria   Brazil   Chile   Czech Republic   France   Greece
Guatemala   Hungary   Italy   Japan   Poland   Portugal   Singapore
South Korea   Switzerland   Thailand   Turkey   Ukraine   Vietnam

Oxford is a registered trademark of Oxford University Press
in the UK and certain other countries.

Published in the United States of America by
Oxford University Press
198 Madison Avenue, New York, NY 10016

Library of Congress Cataloging-in-Publication Data
Barash, David P.
Buddhist biology : ancient Eastern wisdom meets modern Western science / David P. Barash.
pages cm
ISBN 978-0-19-998556-2 (alk. paper)
1. Biology—Philosophy.   2. Biology—Religious aspects.   3. Buddhism and science.   4. Philosophy and
science.   I. Title.
QH331.B244 2013
570.1—dc23
2013007312

9 8 7 6 5 4 3 2 1
Printed in the United States of America
on acid-free paper

To Noa Bogaisky

"If we had a keen vision and feeling for all that is ordinary in life, it would be like hearing the grass grow or the squirrel's heart beat, and we should die of that roar which lies on the other side of silence."

—George Eliot

Contents

Acknowledgments

THERE ARE SO many people (and more than a few animals) who contributed in various ways to this book—not just the writing, but even more, its conceptual foundations—that I'm tempted to thank everyone by thanking no one. Thus, my gratitude is literally unending to a panoply of biologists, Buddhists, philosophers, naturalists, students, and teachers, some long gone and others still flourishing. Special appreciation, however, goes to my agent, Howard Morhaim; to Jeremy Lewis, acquisitions editor at Oxford University Press as well as to eagle-eyed yet light-fingered copy editor, Barbara Norton; and to Tom Kusel, Aditya Ganapathiraju, and James Woelfel, plus a hard-working group of beta-test readers from my Psychology 480 (Ideas of Human Nature) class at the University of Washington. Much of this book was written while on two sabbatical leaves in Costa Rica, for which I thank the Department of Psychology and administration of the University of Washington.

I also wish to note that several different paragraphs, scattered throughout this book, appeared earlier in various essays of mine that were published in *The Chronicle of Higher Education Review*. I am grateful to *The Chronicle* for permission to reprint this material. Similarly, selections from a piece titled "Only Connect," and posted in *aeon* magazine (www.aeonmagazine.com) on Nov. 5, 2012, are incorporated into the present book; I appreciate the kindness of *aeon's* editors in permitting its republication.

BUDDHIST BIOLOGY

1

A Science *Sutra*

EARLY IN SHAKESPEARE'S play *The Tempest*, a clown named Trinculo takes shelter from the storm in a most unappealing place: under the monster, Caliban, explaining that "misery acquaints a man with strange bed-fellows." This phrase has since morphed into "politics makes strange bedfellows." But in fact, there have been many strange bedfellows, not all of them resulting from misery or involving politics.

Prominent among these odd couples is the pairing of religion and science. Which is the clown and which the monster? Maybe both, or neither. Or maybe a bit of each, depending on circumstances. The "new atheists" (notably Richard Dawkins, Daniel Dennett, Christopher Hitchens, and Sam Harris) have claimed that religion and science are not just separate but downright antagonistic. The late Stephen Jay Gould, by contrast, made a case that science and religion (Trinculo and Caliban) are actually compatible because they constitute what he called "NOMA"—Non-Overlapping Magesteria.[1] For Gould, science explains *how* things are, while religion deals with *why*; that is, science is largely concerned with the facts of the world, whereas religion deals with issues of ultimate meaning and ethics. Accordingly, the two are and should be of equal but independent status.

This accommodationist position is appealing, especially since it opens the door to peaceful coexistence between these two key human enterprises. ("I will here shroud till the dregs of the storm be past," concludes Trinculo.) But wishing doesn't make things so, nor does it cause storms to cease. In my opinion, science and religion are often in conflict, not so much because science makes claims about meaning or ethics (although that has happened on occasion), but because religion keeps making assertions about the real world that not only overlap those of science, but are frequently contradicted by them. Consider, for example, the Jewish story about Moses parting the Red Sea and speaking with God in the form of a burning bush; the Christian beliefs that Jesus was born of a virgin, walked on water, raised the dead, and was himself resurrected; or the Muslim insistence that Muhammad took dictation from Allah via the angel Gabriel, and that ten years after becoming a prophet, he traveled to the Seventh Heaven on the back of the "buraq," a white winged steed with a human face as well as an extra set of wings, and whose every stride extended to the farthest point that it could see.

There is, however, an intriguing exception to what I, at least, see as the conflict between science and religion: Buddhism. Perhaps this is because Buddhism is as much a philosophy as a religion, or maybe because Buddhism is somehow more "valid" than, say, the big Abrahamic three (Judaism, Christianity, and Islam). In any event, when it comes to Buddhism and science—especially the science with which I am most familiar, namely biology—we can replace NOMA with "POMA": Productively Overlapping Magisteria. Buddhism is a religion—or, rather, a spiritual and philosophical practice tradition—whereas biology is science. Buddhism is mostly Eastern, at least in its origin; biology is comparably Western. And yet, Kipling was wrong: the twain have met. And not only that, but to a large extent, they get along just fine. Strange bedfellows indeed, yet oddly compatible—at least on occasion, and within limits.

Religion may well be advised to steer clear of factual material, and science, similarly, to avoid deriving "ought" from "is," but the painful truth is that the two have often intersected, with each making "truth claims" that impinge upon the other. And when this has happened, religion has had to retreat…reluctantly, to be sure, and in most cases only after having done much harm to our species-wide search for truth (not to mention bodily harm to many of those pursuing this search). But eventually, religion has given way. The two greatest such fallbacks have involved the geocentric universe, replaced in due course (although not without much gnashing of theological teeth and burning of heretics) by heliocentric reality as described by Nicolaus Copernicus, Johannes Kepler, Galileo Galilei, and Isaac Newton; and the question of special creation as presented in the Judeo-Christian Bible versus evolution, as described by Charles Darwin and supported by decades of modern biological science.

Of these, the second is still a work in progress. The pattern is altogether clear, however: religious fundamentalism, when it impinges on science, has in all cases been fundamentally wrong.[i]

---

[i] Nonetheless, the true believers remain stubbornly unconvinced: For example, in the year 2012, Paul Broun—Republican congressman from Georgia and a member of the House Committee on Science, Space, and Technology, no less—announced that "evolution, embryology and the Big Bang theory—all of that is lies straight from the pit of hell."

To a lesser extent, science has sometimes made claims that are better situated within the realm of religion, or at least, ethics. An especially egregious case has been "evolutionary ethics," the misguided effort to derive moral rules from Darwinian principles; "survival of the fittest," for example, has on occasion been proclaimed not just a guiding principle of the natural world, but a useful arbiter of what *should* be. Such incursions have, however, been mercifully brief and ineffectual: no one, to my knowledge, argues that in deference to the law of gravity, we ought to crawl on our bellies instead of standing upright, or that the Second Law of Thermodynamics mandates that we refrain from making our bed or tidying up the living room.

According to Tenzin Gyatso, better known as the 14th Dalai Lama,

> Suppose that something is definitely proven through scientific investigation, that a certain hypothesis is verified or a certain fact emerges as a result of scientific investigation. And suppose, furthermore, that that fact is incompatible with Buddhist theory. There is no doubt that we must accept the result of the scientific research.[2]

This will be my approach in the present book; namely, that whenever the two come into conflict, science trumps religion every time. Which raises this question: why, bother, then, with any religion—in our case, Buddhism? Maybe pointing to occasional convergences and parallelisms is a foolish enterprise, as one might note a series of random coincidences. On the other hand, maybe there is more to such circumstances than we currently know. What seems clear is that whenever there is uncertainty about anything, each agreement—especially if independently derived—is a separate vote of confidence. And so, the various similarities between biology and Buddhism should give positive pause to anyone who doubts either one. As we'll see, these convergences also promote a worldview that offers not only deeper understanding but also some guides for personal conduct.

One of the Buddha's best-loved stories (he had a lot of them)[ii] concerned a event in which a number of blind men investigated an elephant. The occasion was famously described by the nineteenth-century English poet John Saxe, who wrote:

> It was six men from Industan, to learning much inclined,
> Who went to see the elephant (though all of them were blind),
> that each by observation might satisfy his mind . . .

[ii] Udana 69f.

In the Buddha's version, as well as Saxe's, each felt a different part of the pachyderm, and so the one touching the legs thought it was a tree, the one touching the tail thought it was a snake, and so forth. As it turned out, the blind men

> disputed loud and long, each in his opinion stiff and strong,
> though each was partly in the right, and all of them were wrong.

Buddhism is a bit of an elephant.[iii] The word "Buddhism" didn't even appear in its current Western usage until the 1830s, and a case can be made that there is no such thing as "Buddhism"; rather, just as "Christianity" isn't unitary (there are Roman Catholics, Eastern Orthodox, and Protestants, among others, each of which is further subdivided in proportion as one is familiar with the various denominations), there are many "Buddhisms," notably including the *Theravada* tradition of Southeast Asia (Myanmar, Thailand, Cambodia, Laos, and Sri Lanka), and the *Mahayana* form, practiced primarily in Vietnam as well as in Central and East Asia (China, Tibet, Nepal, Bhutan, Korea, Taiwan, and Japan). In addition, Mahayana separates into a unique Tibetan form as well as the Zen tradition (especially prominent in Japan and Korea) and numerous modified manifestations of Western Buddhism, increasingly popular in the United States and Europe. Among these traditions encompassed by Buddhism, Theravada tends to emphasize personal enlightenment and claims to be closer to the Buddha's original perspective, Mahayana is more focused on assisting one's fellows in their life journey, Zen concerns itself especially with the immediate, practical and down-to earth, and so forth. However, they all share many of the key teachings associated with the Buddha, including concern with the sources of personal pain and disappointment (*dukkha* in Sanskrit), ways of ameliorating *dukkha*, and insights into the nature of reality and of living things.

This book will necessarily be somewhat misleading in that for the most part, I shall consider Buddhism to be a single entity whereas it is multifaceted and complex, an elephant indeed. I trust this is acceptable to all but the most well-informed—and presumably well-intentioned—purists, because I am not trying to provide a detailed introduction to the various schools of Buddhism; lots of other Buddhist adepts, better informed and more competent than I, have already done this. Rather, I shall paint with a broad brush, looking for basic patterns and relationships between biology and Buddhism. Like an elephant, however—in which, after all, a collection of seemingly disconnected parts (trunk, head, body, legs, tail, etc.) really does come

---

[iii] Not to be confused with Ganesh, the much-loved Hindu deity typically represented as a person with an elephant's head. Actually, Hinduism stands in relation to Buddhism somewhat as Judaism does to Christianity: the Buddha was a Hindu before he attracted his own followers, just as Jesus was originally a Jew.

together to form a recognizable if unusual creature—there is something identifiable as Buddhism, at least when viewed from sufficient distance.

For some, Buddhism isn't a religion at all, especially since devotees typically don't accept the existence of an all-powerful creator god who answers prayers and concerns him- or herself with our affairs.[iv] On the other hand, there are Buddhist traditions that identify various supernatural beings as gods, spirits, ghosts, demons, and so forth; indeed, there is also a tendency among some strands of Buddhism to literally worship the Buddha as a god[v]—even though in his teachings Siddhartha Gautama—also known as Sakyamuni, or the Bhagavat ("Lord")—explicitly stated that he wasn't one and shouldn't be treated as if he were in any sense divine.

More than other religions—indeed, I would say, *more than any other religion*—Buddhism lends itself to a dialogue with science. Why? Because among the key aspects of Buddhism, we find insistence that knowledge must be gained through personal experience rather than reliance on the authority of sacred texts or the teachings of avowed masters, because its orientation is empirical rather then theoretical, and because it rejects any conception of absolutes.

On the other hand, Buddhism resembles other religions in many respects. For example, it has followers who identify themselves as such, often by special rituals as well as styles of dress; it offers a compelling personal and emotional dimension to its practitioners; it consists of numerous narratives involving specific events in the lives of its founders, the Buddha most especially; it is summarized in a rich literature consisting of numerous stories and lessons; its followers draw upon a substantial body of ethical teachings and legal constraints; it is institutionalized in an array of social units (the *sangha*), composed of like-minded practitioners; and it boasts a wide variety of physical structures (monasteries, temples, statues) as well as sacred sites and complex works of art. Most important for our purposes, Buddhism offers a well-established doctrine—more precisely, a coherent array of such doctrines—that is less a belief system than a series of postulates claiming to provide insight into the nature of human beings and into the world where they operate.

Buddhism unquestionably offers numerous opportunities for those seeking mystical personal experiences, just as it offers ethical guidelines as well as physical materials to study, admire, collect, and revere, along with the prospect of supportive social networks. In the present book, however, we shall focus on some of the

---

[iv] In the Christian world, a useful distinction can be made between theists and deists: the former believing that God is concerned with the world's day-to-day events, while the latter maintain that, having created the world, God has pretty much left it alone.

[v] "Buddha" is not a personal name but an honorific meaning "enlightened one." Thus, one can speak of "the Buddha" just as Christians speak of "the Christ."

philosophical and doctrinal aspects of this particular elephant—that is, what we might call Buddhism shorn of its abracadabra and hocus-pocus.

Just as it is useful to treat Buddhism as a unitary phenomenon, one among many religious-philosophical-practice traditions—and something consisting, in turn, of a kaleidoscope of diverse subroutines—biology can also be seen as unitary, a coherent discipline comparable to chemistry, physics, geology, and so on. Like Buddhism, however (and like other sciences as well), biology is actually a heterogeneous and slippery thing, consisting in turn of many components: ecology, evolution, genetics, development, microbiology, physiology, anatomy, taxonomy, paleontology, and so forth. Not all aspects of biology are fruitful when compared and contrasted to Buddhist thought. This book will concern itself with those that are. There are many opportunities for the Buddhist and biological perspectives to coincide, and of these I'll focus especially on ecology and evolution, but with more than a passing nod to development, genetics, and neurobiology.

Evolutionary biologists typically attribute convergence among living things not to mere coincidence but to an underlying similarity in ecological niches. If two species—say, whales and sharks—appear very similar despite being unrelated, the likelihood is that both are adapted to a similar marine environment, which of course they are. In the same way, we also have both placental mice and marsupial mice, and the wings of birds and those of bats. The shape and structure of wings speaks eloquently about the physical properties of air and about how the experience of air is shared by flying birds and flying mammals, just as fins are eloquent about water. Likewise, the extraordinary parallels between Buddhism and biology, independently captured by seemingly disparate perspectives, may also be due to the nature of their shared reality.

Full disclosure: I have been a practicing biologist for more than four decades and an aspiring Buddhist (or "Buddhist sympathizer") for about as long, but I am definitely more the former than latter. I have no religious "faith," if faith is taken to mean belief without evidence. Indeed, I have a powerful distrust of organized religion and a deep aversion to anything—*anything*—that smacks of the supernatural. Give me the natural, the real, the material, every time.

In an enterprise such as this book, it is important to be clear about one's starting assumptions. What follows, therefore, is a kind of personal manifesto, critiquing some of the regrettably unscientific aspects of Buddhism—clearing away the superstitious rubble and, I hope, setting the stage for an exploration of common ground between Buddhism and biology.

I am a Buddhist atheist, a phrase that may seem contradictory but that has legitimacy not only in my case, but as a description of many others, of whom the former Buddhist monk and current scholar and author Stephen Batchelor is best-known.

Batchelor worries that as Buddhism has become institutionalized—something that happened quite early in its history—it has "tended to lapse into religiosity."[3] He is also clearly more of a Buddhist than I, and I suspect that I'm more an atheist than he. I further suspect that there are many Buddhist atheists. Moreover, I would bet that Buddhism has a higher proportion of atheists among its devotees than does any other "religion." Thus, Buddhist monks are not believed—even by self-identified devout believers—to possess supernatural powers. Unlike Catholic priests, for example, they are not considered to be intermediaries between God and humanity.[vi] (As noted earlier, one response to this is that maybe Buddhism isn't "really" a religion.)

By contrast, it is hard to imagine a Muslim or Christian atheist, since the terms are oxymoronic: they contradict each other. Someone who self-identifies as a Muslim by definition believes, at minimum, in Allah, in the divine authority of the Qu'ran, in the legitimacy of Muhammad as an inspired prophet, and so forth. Similarly, a "Christian" who doesn't believe in the divinity of Jesus would seem not only a poor Christian but no Christian at all. Interestingly, Jewish atheists are comparatively abundant, probably because unlike Islam and Christianity, whose followers are defined as those who espouse the tenets of their religion, Jews are defined as much by their ethnicity as by their religious beliefs. There are also many "Jew-Boos," people who self-identify both as Jewish and as Buddhist. I would call myself an atheist Jew-Boo: I like matzo-ball soup, tend to talk too much and too loudly, very much value scholarship as well as family, am politically progressive (verging on radical), and I've never been to a NASCAR event.

The conflict between Christianity (and, more recently, Islam) and science—notably evolutionary science—has left many scientists in general and biologists in particular highly skeptical of religion. I am no exception. But maybe we all need to be a bit more open-minded.

There is, nonetheless, much in Buddhism that I reject. Certain practitioners claim, for example, that by virtue of their meditative skill,[vii] they can remember their existence in prior incarnations, or that they can envision their forthcoming

---

vi Tibetan lamas are an exception in this regard; their role among Tibetan Buddhists is roughly analogous to that of priests in the Catholic church.

vii As part of Siddhartha Gautama's claim to have achieved Buddhahood, he announced that he could recall his many thousands of prior lives and that, moreover, he had obtained a "divine eye" as part of his enlightenment, which enabled him to see the karmic consequences of the prior lives of all other sentient beings. This assertion remains potent among most present-day practitioners of Asian Buddhism but is, to me, sheer poppycock—because it not only goes contrary to what we know from science, but is also utterly irrefutable as well as unprovable.

experience after their (next) death. Others maintain that by ritualistically chanting the name of a particular version of the Buddha, they can achieve literal immortality, freedom from pain, cures for life-threatening disease, and so forth.

High on the list of Buddhism absurdities are the phenomena of *iddhi*, supernatural events that are supposed to be generated by extremely skillful and committed meditation. They appear often in Buddhist texts, and I don't believe a word of them. Among the more widespread *iddhi* is the phenomenon whereby enlightened meditators are ostensibly capable of creating illusory duplicates of themselves or of other things. There seems little doubt that such events—the creation of *manomaya*—are themselves illusory.

In addition, higher meditators have been claimed to possess various supernatural abilities, such as becoming invisible on demand; walking through walls, on water, or through the air; being able to hear people and other beings very far away, to mind read, and to recall their past lives; and possessing "divine eyes" that permit them to see the arising and passing away of karma. We are told, as well, that when the Buddha was born, celestial maidens proceeded to scatter flowers from the sky and nine dragons spouted water to bathe the body of the newborn prince. The bodies of monks who die while abiding in a specially blessed meditative state translated as "pure light" are said to remain "uncorrupted"—free from rot.[viii]

By the same token, immediately after he was born, the Buddha ostensibly stood up, took seven steps, and announced that this was the last time he would be reborn. And when, eighty years later, he died, lying on his side between two Sal trees, they immediately and miraculously burst into bloom, despite the fact that they were out of season. Yet the Buddha made it clear that he had no evidence for the existence of a personal soul (*atman*) nor for its more generalized, cosmic counterpart, *brahman*.

More generally, traditional Buddhist cosmology holds that the world is actually flat, with a giant mountain—known as Meru—at its center, and that insects are born from droplets of water. The Dalai Lama (a self-proclaimed admirer of Western science) has recently admitted that he no longer believes that Mount Meru is the center of the world; I don't know about the beliefs of other leading Buddhists in this regard.

The traditional Buddhist cosmology is, however, very specific, and more than a little weird, with the world composed of thirty-one levels. The lowest is a kind of hell, followed in turn by animals, ghosts, titans, humans, five different tiers of lesser gods, fifteen of higher gods, after which one encounters, in turn, "infinite space," "infinite consciousness," "nothingness," and finally "neither perception nor nonperception."

[viii] Similar powers are claimed for certain saints in the Russian Orthodox Church, leading to the unmasking of Father Zossima in Dostoyevsky's *The Brothers Karamazov* as someone less than saintly because Zossima's body did in fact decompose, to the horror of his surviving parishioners.

Of these, the lowest eleven tiers encapsulate the sphere of sense desires, the next fifteen constitute the "sphere of pure form," and the final four, the sphere of formlessness."

The historian of religion David McMahan describes what he calls "Buddhist modernism" as departing from traditional, originalist Buddhism in several respects, one of the most important for our purposes being "demythologization," in which traditional deities and practices are de-emphasized—most often, ignored altogether.[4] For example, so-called hungry ghosts (*pretas*) are traditionally thought to have arisen because of the greedy karma accumulated via their past lives; they are pictured as having skinny, pencil-thin necks, bloated bellies, and incessant, unsatisfied hunger. In traditional Buddhist practice, food is ceremonially offered to them so as to buy their goodwill, thereby preventing them from doing harm.

These days, especially in the West, hungry ghosts are mostly considered (when they are considered at all) to be unconscious manifestations of one's own dissatisfactions and neediness; that is, they have largely been reinterpreted as psychological entities rather than genuine ones. This is another aspect of Buddhist modernism: reinterpreting what had been taken literally as reflecting instead psychological but not physical truths.

Similarly, according to the *Tibetan Book of the Dead* (*Bar Do Thos Grol*), deceased people are said to pass through three distinct stages (*bar dos*) between their demise and their next reincarnation. Among Tibetan Buddhists, these intermediate stages are populated by various Buddha images, some peaceful and serene, others threatening and wrathful. For Buddhist modernists—notably Carl Jung and his followers, but including non-Jungians as well—these encounters reflect various deep-seated psychological archetypes, that is, our own internal psychic forces, rather than "real" entities. Traditional Buddhists would disagree.

As McMahan writes,

> For traditional Himalayan Buddhists, the world is alive not only with awakened beings but also countless ghosts, spirits, demons and protector deities. These beings are prayed to and propitiated in daily rituals and cyclical festivals, and they figure into one's everyday life in very concrete ways.

For a dramatic and tragic example, a controversy known as the "Shukden affair" erupted within the Tibetan Buddhist community beginning in the late 1970s. The fourteenth (and current) Dalai Lama had aligned himself against more traditional, conservative members of his particular variant of Buddhism known as the Gelukpa school. At issue was whether a particular deity known as Dorje Shukden should or should not be propitiated via certain rituals,[ix] although both sides agreed that Shukden was "real." Things

---

[ix] Interestingly, this was *not* a simple matter of the Dalai Lama's advocating rationality as against superstitious traditionalists. Thus, the Dalai Lama (who opposed making offerings to Dorje Shukden) nonetheless considered

grew so heated that a supporter of the Dalai Lama's viewpoint was murdered in 1997, along with two of his students.

It should be emphasized, however, that unlike the major schisms between, say, Protestants and Catholics, there haven't been any persistent, violent clashes among competing factions, with one branch accusing the other of heresy. Nothing has occurred even remotely like the St. Bartholomew's Day Massacre, and there have been no cases of murderous enmity between Mahayana and Theravada Buddhists comparable to the centuries of "troubles" between Irish Catholics and Protestants.

A final example in which I (and many other Buddhist sympathizers) part company with traditional Buddhist beliefs concerns the doctrine of reincarnation. As we shall see, there is a very limited sense in which reincarnation can in fact be interpreted as consistent with modern biological science, but definitely not in the conventional sense of Buddhism or Hinduism—that is, in which individuals as opposed to their constituent molecules are somehow reconstituted in their characteristic personalities. For those of us interested in reconciling Buddhism with science in general and biology in particular, traditional reincarnation remains a pronounced and irreconcilable outlier.

The idea of reincarnation nonetheless persists in the minds of many people, perhaps especially those who cannot abide the notion of their personal death but who also resist the standard Western religious imaginings of a literal heaven and hell. There is something downright grotesque, however, about literal reincarnation, beautifully caricatured by Vladimir Nabokov in his wildly imaginative novel *Pale Fire*. At one point, the author wonders "How not to panic when you're made a ghost," specifically, being faced with the prospect of "freak reincarnation." He asks

> …What to do
> On suddenly discovering that you
> Are now a young and vulnerable toad
> Plump in the middle of a busy road…

This is a misunderstanding of reincarnation, Buddhist or otherwise, which normally presupposes that the soul or spirit to be reincarnated isn't suddenly inserted into an existing body but rather is somehow introduced into a new and developing one, when that body is born (or, in most Buddhist traditions, at the time of conception). But Nabokov's little jingle helps emphasize the silliness of the concept.

---

him to be a "spirit of the dark forces" and objected to the Shukden rituals because he worried that they would hurt the cause of Tibetan unity as well as his own personal health.

Nonetheless, as with reincarnation, many people are especially tempted to suspend some of their critical faculties when dealing with Buddhism and other Eastern traditions, partly because they are appealingly exotic. This is especially true of open-minded Westerners who view their own dominant culture with distrust and alienation. But such tolerance verging on admiration is not a route that I can follow, or recommend. I simply will not credit any faith-based assertions that defy rationality and empirical scientific fact.

We would also be well-advised, I believe, to acknowledge and avoid the danger of announcing false parallels, which sometimes reflects inappropriate enthusiasm on the part of those seeking to reconcile religion and science. For example, did not Nicholas of Cusa write in *De Visione Dei* that "the wall of the Paradise in which Thou, Lord, dwellest is built of contradictories"? And isn't this pretty much like the dual particle-wave nature of light, which twentieth-century physicists have confirmed? And did not Dionysius the Areopagite say in *The Divine Names* that "He is both at rest and in motion, and yet is in neither state," thereby anticipating Heisenberg's indeterminacy principle? I am not making up these examples; they appear in the writing of a highly regarded expert on world religions and represent a kind of irrational exuberance that—for all its temptations—I shall try to avoid.[5]

Turning specifically to Buddhism, the present book will likely trouble those otherwise gentle Buddhist souls who so revere Tenzin Gyatso that they append to his name the honorific "HH," His Holiness. "The Dalai Lama" is okay with me, since that is how this particular gentleman is widely known, but even though I greatly admire him for his kindness as well as his wisdom, I cannot swallow the notion that he is any holier than thou, or me, or Charles Darwin, or anyone else. Either we are all holy (whatever that means), or no one is.

One of the historical Buddha's most appealing—and biologically valid—teachings is that everyone possesses "Buddha nature."[x] We'll see that as a statement of biological continuity among all living things, this teaching is not only admirable, it is also scientifically accurate. But as a proclamation of divinity, it is more than a little suspect.

Moreover, the claim that Gyatso is literally the fourteenth incarnation of the original Dalai Lama is, I firmly believe, nothing but arrant nonsense, comparable to the wacky Mormon story in which Joseph Smith allegedly discovered God's words inscribed on some golden plates in Elmira, New York, or the bizarre assertion that Moses had a personal audience with a deity who presented him with an inscription,

---

[x] Stephen Batchelor tells me that this concept as literally stated was never used by the historical Buddha; it first appears about a thousand years later and is based on the view that all beings are capable of attaining awakening.

written in stone, detailing what the Israelites should and should not do, or that Jesus of Nazareth was born of a virgin, raised the dead, and changed water into wine.

In short, if you, dear reader, are looking in these pages for obeisance to anything that smells even remotely like traditional religious dogma or spiritualist mumbo-jumbo, you won't find it. By the same token, I shall do my best to avoid distorting either Buddhism or biology in my enthusiasm for identifying parallels and convergences between the two.

Of course, you can today find literally millions of people who identify themselves as "Buddhists," hanging out in temples—especially in Asia—and praying to one or more idols identified as "Buddha," believing in the literal existence of a Buddhist deity and requesting the same sort of divine intervention that many Christians, Jews, or Muslims seek when importuning their personal god for one thing or another. This speaks volumes about certain widespread human needs, but it says nothing whatever about truth, human or otherwise, or about science.

I hold to the position that Buddhism in its most useful, user-friendly, and indeed meaningful form is not in fact a religion in the standard Western sense of the term. Rather, it is a perspective, a philosophical tradition of inquiry and wisdom, a way of looking at the world that is often perverted into a kind of "sky-god" faith complete with other nonsensical rigmarole, but, in its more genuine form, is anything but. This view has been superbly developed by Stephen Batchelor in his aptly titled *Buddhism without Beliefs*. I won't revisit that ground except to add an ironic "Amen."

The basic idea of Buddhist biology—the phenomenon as well as this book—is straightforward enough: Buddhism and biology have a great deal in common, such that tremendous insights into each can be gained by learning about the other. There is a special charm in attempting to reconcile apparent opposites, such as particle and wave, matter and energy, yin and yang. And yet, "Buddhist biology" might threaten to be just a meaningless verbal confluence, an imaginary entity whose pairing owes more to alliteration than to reality (why not "Confucian chemistry"? or "Jewish geology"?). I hope to convince you instead that Buddhism—not the abracadabra religion, but concepts based on the ancient Eastern philosophical and practice tradition—enjoys remarkable parallels with at least one aspect of modern, Western, hardheaded science.

In his book, *Physics and Philosophy*, Werner Heisenberg – one of the commanding figures in the development of modern quantum physics – wrote that

in the history of human thinking the most fruitful developments frequently take place at those points where two different lines of thought meet. These lines may have their roots in quite different parts of human culture, in different times, different cultural environments or different religious traditions: hence if they actually meet, that is, if they are at least so much related to each other that a real interaction can take place, then one may hope that new and interesting developments may occur.[6]

I am quite sure that Buddhism need not imply mysticism (any more than physics necessarily does).[xi] In fact, the opposite is more reliably true, insofar as Buddhism, in its most powerful and useful form, is practical, immediate, and down-to-earth. Most laypersons, however, believe otherwise, classing Buddhism along with various other "Eastern mystical" traditions. One respect in which Buddhism and mysticism do in fact converge is when it comes to the role of personal experience: both treasure it.

And here is an interesting and counterintuitive fact: this aspect of "mysticism," namely the primacy of the subjective, also makes science possible. Mystics have long maintained that the only valid guide to truth is each individual's personal experience, while science, too, requires direct experience, namely personal encounters with the natural world, which leads to the generation of hypotheses followed by empirical testing and either confirmation or refutation.[xii]

Science, however, isn't simply a matter of personal experience; it relies on verifiability, the basic notion that in order for something to be accepted as true, it must exist in the real world. Accordingly, it must be more than anyone's personal, subjective (mystical?) truth. Nonetheless, all encounters with reality begin with just that: the experience of an encounter, which by definition is subjective.

I am not arguing that science is literally mystical, but rather that what passes for mystical can—at least in some circumstances—be a gateway to scientific understanding. Moreover, this, in turn, is especially true when it comes to Buddhism, which—more than any other wisdom tradition—embraces the importance of personal, subjective, empirically verifiable experience.

The comfortable fit between Buddhism and empirical science has been facilitated by several teachings, of which one of the most important is the Kalama Sutra.[xiii] In it, the Buddha advises his audience (people known as the Kalamas) how to deal with the bewildering diversity of conflicting claims on the part of various Brahmans and itinerant monks:

Do not go upon what has been acquired by repeated hearing; nor upon tradition; nor upon rumor; nor upon what is in a scripture; nor upon surmise;

[xi] Although some people have made a career—or at least, a splash—by connecting physics with Eastern philosophy.

[xii] According to the philosopher Karl Popper, scientific hypotheses can never be altogether proved; instead, the hallmark of science is that its ideas lend themselves to being *disproved*.

[xiii] Since this teaching appears only in the original Pali Canon, it should strictly speaking be identified as a *sutta*, the Pali transliteration; however, for consistency I am using only Sanskrit-to-English spelling in this book.

nor upon an axiom; nor upon specious reasoning; nor upon a bias towards a notion that has been pondered over; nor upon another's seeming ability; nor upon the consideration, "The monk is our teacher." Kalamas, when you yourselves know: "These things are good; these things are not blamable; these things are praised by the wise; undertaken and observed, these things lead to benefit and happiness," enter on and abide in them.[7]

This teaching is widely—and appropriately—seen as supporting free inquiry and an absence of rigid dogma, an attitude entirely open to empirical verification and thus consistent with science. Nonetheless, a case can also be made that the Kalama Sutra isn't quite so data oriented and thus empirically open-minded and pro-science as claimed. For one thing, those being exhorted are not the Buddha's own disciples, urged to treat skeptically the Buddha's own teaching, but people who had been following other teachings (notably, various branches of Hinduism) and who therefore might well have been considered ripe for the plucking—something like pagans being urged by a Christian missionary to abandon their old ways. It is the rare evangelist who, in urging his listeners to question their prior, heathen beliefs, insists just as firmly that they also question the new ones that he is promoting.[xiv]

In addition, the Kalama Sutra appears to be primarily focused on moral teachings rather than matters of empirical fact and subsequent belief. Worth noting as well is that until recently such skepticism was not generally considered central to Buddhist teaching; that has come more recently. On the other hand, the words of the Kalama Sutra are quite straightforward, and they do in fact fit nicely into the Western scientific tradition: the Royal Society of London, whose full name was the Royal Society of London for the Improvement of Natural Knowledge, and which was the world's first—and for a long time, its foremost—scientific society, has as its credo *Nullius in verba*—"On the words of no one."[xv]

Returning once again to the Buddha's emphasis on validation by experience rather than via hierarchical or scriptural authority, consider this statement, which could as well have been uttered by a senior Nobel-winning scientist, advising junior researchers in his laboratory: "Just as one would examine gold through burning, cutting, and rubbing so should monks and scholars examine my words. Only thus should they be accepted, but not merely out of respect for me."[8] On balance, it seems reasonable and appropriate that Buddhism be viewed in the West as comparatively free of

---

[xiv] But then, orthodox Buddhists would doubtless argue that the Buddha was a rare one indeed (and they would probably be correct).

[xv] In addition, troublemakers might ask about the "authority" by which the Buddha exhorted others not to accept teachings by virtue of the teacher's authority…but that's another matter. (By the same token, one might ask by what authority the Royal Society insists that its own credo be followed.)

irrationality, superstitious belief, and stultifying tradition—but this generalization must nonetheless be taken with a grain of salt, noting that in much of the world, Buddhism involves daily ritual devotions, belief in amulets and other special charms, and even the presupposition that Siddhartha Gautama was not a man but a divine being.

This book is neither a recounting of Buddhist history nor a detailed exegesis of Buddhist philosophy, and it certainly won't give you advice on how to meditate or how to become a better, happier, or more centered, compassionate and balanced person. Insofar as it offers enlightenment—and I would like to think that to some extent it does—it is purely of the cognitive dimension: information, analysis, and intellectual accomplishment of the sort that comes when relationships are clarified, ideas explained, and, on occasion, new insights gained.

Buddhist meditation is often understood as equivalent to Christian, Islamic, or Judaic prayer, but whereas prayer typically requests God to intervene in some manner in human affairs, Buddhist meditators aim simply to purify their minds and to cultivate relaxation, compassion, and wisdom. There are basically two forms of meditation: *samatha* and *vipassana*, which are close enough to each other that to some extent any distinction is arbitrary. However, *samatha* places particular emphasis on calming and relaxing the meditator's mind, whereas *vipassana* is especially concerned with insight. In this regard, the current book is something of an extended *vipassana* meditation, intended to yield insight. As for inner peace, that's up to you. At the same time, there is some reason to believe that the reader may benefit directly from at least some of the insights that I am attempting to share in these pages. Although we shall spend considerable time examining *dukkha*, my own experience has been that a deeper understanding of Buddhism, especially how it speaks to biology and vice versa, contributes to the development of the alternative to *dukkha*, namely *sukha*. This desirable condition has been described by the former molecular biologist and now longtime Buddhist monk and scholar Matthieu Ricard as

> the state of lasting well-being that manifests itself when we have freed ourselves of mental blindness and afflictive emotions. It is also the wisdom that allows us to see the world as it is, without veils or distortions. It is, finally, the joy of moving toward inner freedom and the loving-kindness that radiates toward others.[9]

And so, dear reader, let us hope that your exploration of "Buddhist biology and vice versa" will provide at least some glimmerings of *sukha*.

In one of the many charming myths surrounding the historical Buddha, it is said that after he gained awakening, he sat beneath the fabled bodhi tree for seven days, contemplating his achievement, whereupon the gods concluded that perhaps Gautama Buddha was less than fully enlightened, since he was still a prisoner of "attachment"—namely, to the circumstances of his own awakening.[xvi] Just to show how wrong they were, the Buddha allegedly performed the "miracle of the pairs," in which he emitted paired streams of water and fire from every pore in his body. Later he brought forth light rays that were blue, maroon, red, white, yellow, and (perhaps to prove that he was an equal-opportunity emitter) clear.

I do not anticipate that anyone reading this book will respond similarly, although Oxford University Press takes this opportunity to note that it expects to be fully indemnified in the event of any unanticipated post-enlightenment emissions, whether fiery, watery, visible, odoriferous, electromagnetic, radioactive, and so on. In short: proceed at your own risk.

Early in my career, I employed short essay exams in an undergraduate class I taught at the University of Washington. One of those now-extinct exams asked students to explain, briefly, Darwin's primary scientific contribution. I still remember one student's answer: "He invented evolution."

Sorry, no cigar...and no credit. (The correct answer, by the way, isn't even that Darwin *discovered* evolution or that he presented abundant evidence in its favor; rather, he came up with the most plausible explanation for the *mechanism* whereby evolution proceeds: namely, natural selection. Others, such as Robert Chambers and Darwin's own grandfather, Erasmus Darwin, had preceded him in describing some sort of historical, evolutionary connection among organisms.) In any event, the student's response helps clarify the difference between inventing or creating something, on the one hand, and discovering or revealing it, on the other, regardless of the mechanism employed.

Whereas the Abrahamic religions—Judaism, Christianity, and Islam—claim that their god literally created the world and along with it the natural laws that govern its functioning, Buddhism promotes a very different perspective, namely that the Buddha—emphatically not a god, by his own insistence—didn't create the *dharma* (the way the world wags); rather, he discovered it. Thus, for Buddhists, reality exists prior to any supernatural event; for the Big Three, it exists only because of just such an event.

To me, at least, this distinction seems important, although I'm not at all sure where, precisely, it leads. Unlike the Abrahamic triad, Buddhism encourages participants to explore for themselves, explicitly enjoining devotees to reject any teachings

---

[xvi] Various Buddhist traditions believe in the existence of numerous gods, most of whom are closer to the Western concept of demons.

that seem incongruent with their own experience of reality. There can and should be a world of difference between believing in a creator god versus a discoverer dude when it comes to interacting with physical reality. Thus, a persistent theme in post-Enlightenment Western thought has been that we can learn about the nature of God the creator by discovering as much as possible about what he hath wrought.

In 1829, Francis Henry Egerton, the eighth Earl of Bridgewater, bequeathed 8,000 pounds sterling to the Royal Society of London to support the publication of works "On the Power, Wisdom, and Goodness of God, as Manifested in the Creation." The resulting *Bridgewater Treatises*, published between 1833 and 1840, were classic statements of "natural theology," seeking to demonstrate, admire, and worship the presumed creator by examining, in detail, his creation. But a problem nonetheless remained, since the rules of that creation are necessarily assumed to be inviolate and perfect, which is problematic when we consider how much imperfection there is. Consider, for example, the blind spot in the vertebrate retina, the lousy design of the human lower back, or the downright ludicrous structure of the urinary and reproductive systems (especially in men). These are all testimony to the ramshackle nature of organisms, whose evolution is necessarily constrained by their preceding histories, leading to outcomes that often belie the assumption that they were in any way intelligently designed.

As it happens, tension between science and religious belief seems to have been even more important in Islam than in Christianity. Thus, roughly a thousand years ago, the Sufi philosopher Al-Ghazzali—whose writing was and still is highly influential in the Muslim world—argued strenuously against anything even approximating a law of nature, since this would by definition restrict the freedom of an all-powerful deity. Al-Ghazzali famously wrote, for example, that when a piece of paper (or maybe it was a ball of cotton) was heated sufficiently, it changed color and gave off heat, flame, and smoke not because it was burning according to its nature, but because it pleased Allah for this kind of transformation to take place at this particular time. Had Allah been of a different mind at such a moment, the fuel would have turned green, remained unaffected, or been transmuted into a pot of tea, a mountain, or a giant ox—whatever Allah willed, independent of any laws of nature or rules of science. Rules, schmules! Laws, schmaws!

For Al-Ghazzali, and for generations of Islamic thinkers following him, it was simply unacceptable for any laws of nature to exist, because they would limit God's options. To a degree, this parallels the traditional Catholic view of the Pelagian heresy, which had claimed that people could secure for themselves a place in heaven by virtue of their good deeds; the problem with this from the pope's perspective was that it suggested that mere mortals could twist God's arm and achieve their own ends in response to their personal desires and chosen actions, whereas in truth, such "decisions" must be up to God alone.

By contrast, Buddhism promotes a worldview in which the *dharma* simply exists[xvii] — gravity, strong and weak forces, photons and electrons and Higgs bosons, the Second Law of Thermodynamics, the phenomenon of natural selection, and, of course, the laws of karma—such that our job is to reveal and understand these realities, without worrying that, like Galileo or Darwin, we might run afoul of prior assertions of God's will as revealed in his perfect and immutable creation. This in turn ought to lend itself to a more liberated, exploratory, and productive approach to understanding all that is, imperfect and unplanned as it may be.

The Buddha's reported experience was not an encounter with God or even with any of his or her messengers. Rather, it is a human being's experience of the world, and is therefore much more amenable to science. On the other hand, there are also claims that simply fly in the face of any perspective that even approaches a scientific one. Some of these, as we shall see, are more properly relegated to the realm of folk tale and fanciful mythology, and are neither more nor less plausible than many other claims of Western religions, such as Noah taking pairs of all animals (and presumably, plants, as well as a full load of very nasty viruses) on board his ark, Moses parting the waters of the Red Sea, or Jesus walking on water feeding a multitude on three little fishes and five loaves of bread or or being resurrected after his torture and execution.

Here is the renowned scholar of Tibetan Buddhism Robert Thurman:

> No matter how preposterous it may seem to us at first, it is necessary to acknowledge the Buddha's claim of the attainment of omniscience in enlightenment. It is foundational for every form of Buddhism. It is rarely brought to the fore nowadays, even by Buddhist writers, since this claim by a being once human is uttermost, damnable sacrilege for traditional theists and a primitive fantasy, an utter impossibility, for modern materialist. But it is indispensable for Buddhists.[10]

I don't wish to be unkind, either to Thurman or to any Buddhas past or future, and certainly not to the legions of currently devout and gentle followers of the Buddha's teachings,[xviii] but as I have already noted, such a claim really is "preposterous," a "primitive fantasy," and "an utter impossibility." This doesn't mean, however, that

---

[xvii] *Dharma* here being Buddha's teaching, but also natural law in its widest and deepest sense.

[xviii] I hope it does not violate any expectation of anonymity if I note that when he reviewed an earlier version of the current manuscript for a different publisher, Thurman was outraged by my intention to unhesitatingly privilege science over religion.

it—or the rest of Buddhist teaching—should similarly be discarded. In this book, I hope to biologize the Buddha, pointing respectfully as a biologist to the useful and insightful teachings of Buddhism, once shorn of its magical trappings. In his book *Alone with Others: An Existential Approach to Buddhism*, Stephen Batchelor notes the tension among modern Buddhists between clinging to "historical and cultural factors" that are no longer relevant, therefore succumbing "to the ossifying forces of institutionalization and formalism," on the one hand, and, on the other, going too far in the opposite direction and thereby "dissolving all distinctions into a nebulous eclecticism, in which all ideas, however tenuous their associations, are magnanimously housed under one roof in beatific ignorance of their incompatibilities."[11]

My goal is something analogous to the Jefferson Bible, which refers to Thomas Jefferson's amendation of the Bible so as to retain what he saw as its powerful insights, once the supernatural nonsense was edited out. When it comes to Buddhism, however, such a project is easier said than done, if only because Buddhism lacks a single, readily identified canon of liturgically agreed-upon sacred writings, analogous to the Old and New Testaments. Like Mark Twain, who said it was easy to stop smoking—he had done it hundreds of times—it is easy to locate Buddhist texts: there are gazillions of them.

Here is Thurman once more writing about the "vastness" of the available literature:

A nation that probably never numbered more than ten million—nowadays six or seven million—has two different canons of Indian texts translated from Sanskrit, numbering over three hundred volumes, each volume of which would translate into a roughly two-thousand-page English text. Radiating out from that canon are collected works in Tibetan of hundreds of eminent scholars, saints, and sages, some of which number in the hundreds of volumes.

And this is just the Tibetan version. After originating in India, Buddhism subsequently spread to Sri Lanka, and then to southeast Asia (Thailand, Burma, Vietnam, and other lands), to China, and also to Korea and Japan—and, more recently, to the West. The reasons for this spread are doubtless complex and multifaceted. Worth noting, however, and relevant to the goal of the present book is the widespread hope that Buddhist wisdom will not only help to humanize science, but also to solve some of the many problems of modernity itself, notably but not limited to environmental destruction, social inequity, violence, excessive consumerism, widespread alienation, and war.[xix]

[xix] Granted, a tall order!

Among the many commonalities between Buddhism and biology, probably the most important is their shared recognition of interconnection between organism and environment. This principle (*pratitya-samutpada* in Sanskrit) is usually translated as "dependent co-origination," "dependent arising," or "interdependent origination." In more modern usage, it underlies ecological insights into the interdependent nature of ecosystems, communities, food webs, and trophic levels as well as the artificiality of separating any living creature from its surroundings.

Crucially underlying this realization, in turn, is the Buddhist concept of *anatman* or "not-self" along with *anitya* or "impermanence." Taken together, these three—*anatman*, *anitya*, and *pratitya-samutpada*—illuminate why compassion is one of the organizing principles of Buddhism: after all, if selves are not really solitary and separate, then compassion is not only appropriate, but unavoidable. Most of us do not exhort our kidneys to be compassionate toward our lungs. It's the way we are constructed. One of the crucial discoveries of evolutionary biology has involved the significance of genetic connectedness, not only between organisms but also within each living thing: it's the way they are constructed. Specifically, body parts cooperate because at the genetic level, they are deeply and fundamentally the same. More on this later.

Similarly, we'll see numerous other parallels between evolutionary biology and Buddhism, including, for instance, an embrace of impermanence. Evolution, for example, is the science of natural, organic change; it points not only to the adaptedness of living things "as they are," but also to their transience. So does Buddhism. We'll explore, as well, the shared Buddhism-biology distrust of simple cause-and-effect analysis, a recognition of life as transient and a process rather than permanent and static, as well as an appreciation of the rightness of nature and of the suffering that results when natural processes are tampered with (and even when they are not).

Buddhism offers, especially to many in the West, an appealing aura of Eastern exoticism combined with the hope of transcending some of the unattractive aspects of modern, high-tech living, with its growing alienation from the natural world as well as from each other. Not only that, but in promising an apparent way out, Buddhism has the added advantage of letting science in.

At first glance, it might appear that Buddhism is irretrievably anti-science because of its refusal to embrace clear-cut alternatives. After all, certain basic distinctions, such as "true or false" seem fundamental to any scientific worldview. And yet, many—perhaps most—scientific disciplines join ecology and evolutionary biology in pointing out the futility of sharply defined either-or dichotomies.

Such a stance has in fact become characteristic of nearly all branches of modern science, including ethology (learning/instinct), genetics (genotype/phenotype), cellular biology (proteins/nucleic acids), morphology (structure/function), physiology (chemical/electrical), chemistry (ionic/covalent), and even that bastion of the "definite," physics (particle/wave, matter/energy). We'll steer clear, however, of physics, not because it isn't a suitable avenue for exploration, but simply because it is a path already well trodden.

For the most part, we'll also refrain from lengthy excursions into "wisdom traditions" (that is, religions) that non-Buddhist Westerners find more familiar. This is not because Buddhism is the only spiritual perspective with implications for our relationship to the natural world, although it is, in my judgment, without parallel as the one whose implications are the most convivial, life-affirming, and—in a word—accurate. For the most part, Western traditions generally fall prey to the same dichotomous dynamic that characterizes much of science: in place of the scientific view that "I am here, the world is there," we typically encounter the religious view that "I am here, God is there."

For good reasons, therefore, Judaism, Christianity, and Islam have been criticized for supporting a bifurcated and often hostile perspective toward the relationship of people to their environment. As we'll see later, there is reason to believe that the book of Genesis constituted a foundation text for this attitude on the part of the Abrahamic big three whereby the Western world received its marching orders, which unfortunately specified a separation between human beings and the rest of "creation." In addition, the Old Testament urged us in the monotheistic West to go forth and multiply while guaranteeing us dominion over all other living things. The result was also to guarantee a lot of headaches, heartache, and a profoundly dangerous, misleading, and hurtful perspective on our biological being, including our place in nature. (More of this in chapter 4.)

For now, let's note that part of the difficulty may well come from monotheism itself. In the West, monotheism stands pretty much unquestioned. Generations of schoolchildren have been taught, for example, to associate different ancient societies with particular accomplishments. China? Gunpowder, hierarchical government, and the Great Wall. The Aztecs? Hot chocolate, popcorn, mandatory education, and the canoe. Egypt? Paper and monotheism. Not quite the right deity, to be sure, but at least the right idea when it comes to worshipping one single solitary overarching omniscient god. As a child, I didn't question this piece of truth, and neither, I suspect, do most children—or adults—today. And maybe, of course, it (i.e., monotheism) really is true. Maybe there really is one and only one God, in which case Akhenaten and his followers genuinely were onto something big. Just for the fun of it, however, let's put that possibility aside and ask: other than the chance that it might be true, what is to be said for monotheism? In what sense—if at all—is it an advance over, say, pantheism (the belief in many gods) or atheism (belief in none)?

For one thing, it's easier to keep track of one than of many. Maybe it isn't easier to follow his or her precepts and rules, but it's easier to keep them straight. There's another practical benefit or two as well. Conceiving a single, unitary deity probably makes it easier to unite disparate people, with everyone rallying round one flag rather than many, singing one song in one voice instead of the discordance of multiplicity. Moreover, if, as Sigmund Freud suggested in *The Future of an Illusion*, religion is an "infantile neurosis" by which people meet their need for a benevolent, all-knowing, and all-powerful parent, it is presumably more satisfying, psychologically, to have one such parent than a whole slew of them.

On the other hand, one god implies—although it does not necessarily require—one truth. And I fear that such conceptual monism leads readily to a degree of intolerance that exceeds that of pantheistic societies, for whom many gods may well augur the existence of many truths. Beyond this, monotheism has implications for how people interpret the workings of the world, especially if there is a holy text, a blueprint of his or her thoughts, plans, and words. It can be argued that Western science has profited from this orientation insofar as the scientific enterprise is indeed a matter of understanding what is seen to be a unitary deep structure of reality, tracing the belts and pulleys that keep everything moving, including the grease that lubricates the gears and the design principles by which everything operates.

It has worked. Centuries of scientific advances have shown that there are certain basic principles by which the world operates: gravity, entropy, electromagnetic force, relativity, natural selection, and so forth. Maybe if we occupied an intellectual universe in which there was a separate god immanent in each bush, rock, tree, and bird, we would be convinced that each entity also acted according to its own separate natural laws, and so science itself would have been particularized, localized—and irretrievably stunted.

But what happens when reality itself can only be apprehended by a more nuanced approach? When the simple certainties of one god and one truth threaten to mislead? I am not thinking here of those obvious cases in which religious dogma is simply wrong, such as Galileo being forced to recant his observations proving that the Earth moves, or those tens of millions of Americans who, even today, prefer blinkered ignorance to the fact of evolution. My point, instead, involves situations in which the monotheistic mind-set leads us—often unknowingly—into error. In such cases, it may be worthwhile for all of us, not just scientists, to consider other intellectual frameworks. Of these, Buddhism offers one of the most congenial, not only in that it provides a paradigm in which interaction, change, interdependence, and a genuine blurring of boundaries is permissible, but because it promotes and encourages such thinking.

A Buddhist outlook is also unusually receptive to paradox, most notably in the famous koans of Zen ("what is the sound of one hand clapping?," "what was your face before you were born?," and so forth). Buddhism generally, moreover—and not just its Zen manifestation—is especially paradox-friendly in its insistence upon *anatman* ("How can I lack a self?") and *anitya* ("Aren't I the same person I was yesterday, and will be tomorrow?"), of which more later. For now, let's note that paradox looms large in evolutionary biology as well. As George Levine notes, in a Darwinian world

[w]hat is stable is in motion; what is enormous depends on minutiae; what seems peaceful is at war; struggle is often mutual dependency; lowly worms create the large green expanses of England; six thousand years is no time at all;...if unchecked by natural selection, even slow-breeding elephants would entirely cover the earth within five centuries; there are woodpeckers living where not a tree grows; there are web-footed birds that never go near the water; we are related physiologically to all living things, not only apes but to barnacles and spiders....The world of brilliant adaptation is moved not by a creative intelligence but by "unknown laws of nature," and in nature itself the only intelligence is that of organisms, and most obviously and particularly humans.[12]

In my own writing and lecturing on evolution and other aspects of biology, I have consistently found Eastern audiences to be more comfortable and even welcoming than those in the West. I don't believe that Asians are smarter, inherently friendlier, or better educated than their occidental counterparts—at least, not in the strict sense. But in one regard they are in fact better situated to absorb certain truths from the natural sciences, notably ecology and evolutionary biology: to some extent, a Buddhist perspective—often creatively combined with that of Hinduism, Taoism, and/or Shintoism—generates far greater openness to the interconnectedness of all life.

For the most part, however, Buddhists have historically been less concerned with explaining the world than with generating personal peace and enlightenment. In chapters 5 and 6, and to some extent chapter 7 as well, we shall look into the most significant departure from this somewhat "selfish" tradition when we examine the growing movement toward "engaged Buddhism." There is a famous *sutra*—discourse attributed to the Buddha—in which a monk named Malunkyaputta comes to the Buddha with ten "unexplained" questions, such as "Is the world eternal or is it temporally limited?," "Is the world finite or infinite?," "Is the soul one thing and the body another?," and so forth. The monk is insistent and agitated—and evidently rather annoying—adding that he cannot continue his spiritual training unless and

until he has obtained satisfactory answers. The Buddha replies with the famous
Story of the Poisoned Arrow:

> There was a man who had been hit by a poisoned arrow and was suffering.
> Although his kinsmen and friends urged him to promptly see a physician, the one
> who really counted, the man himself, asked: "Was the person who shot me with
> the poisoned arrow a Brahman, a commoner, or a manservant? And what was his
> name? Was the person tall or short? What was the hue of his skin? Where does
> he live? I cannot have the arrow extracted until I know these things." The man
> furthered questioned and argued: "As for the bow used for this poisoned arrow,
> what kind was it? What material was the bowstring made from? And what about
> the shaft of the arrow, what kind of feathers were used? What about the type of
> poison?" While he went on in this way, it is said, the poison coursed throughout
> his whole body, and the man unfortunately died.

The lesson? Avoid metaphysics. Keep it real!

And reality is precisely what Buddhism is mostly about. Much of that reality is
neither more nor less than the natural reality of the physical world, especially nature
itself. Consider this foundation tale, told about Mahakashyapa, one of the major
early disciples of the Buddha. The story goes that once, when the Buddha simply
held up a flower as his "*dharma* talk,"[xx] only Mahakashyapa "got it." He responded
by smiling. Mahakashyapa has subsequently been widely acknowledged as the first
Indian founder of what eventually became Zen Buddhism.

Then there is this poem, titled "At Home in the Mountains,"[13] written nearly one
thousand years ago and representative of a continuing tradition:

> Strung with garlands of flowering vines,
> This patch of earth delights the mind;
> The lovely calls of elephants sound—
> These rocky crags do please me so!…
> Not occupied by village folk,
> But visited by herds of deer;
> Strewn with flocks of various
> These rocky crags do please me so!

[xx] "*Dharma* talk" is a structured lecture, roughly equivalent to a sermon, attributed to the Buddha; it remains a
part of Buddhist educational outreach to this day.

More than any Western religious perspective, Buddhism has long been deeply appreciative of and involved with the natural world.[xxi] This is especially true of its East Asian manifestations, generated when classical Buddhism found itself exported to China, where it mixed smoothly with preexisting Taoist perspectives, which had long practiced immersion in the sights, sounds, smells, and textures of untrammeled nature. The Buddha himself, moreover, is said to have achieved enlightenment while meditating under a tree, just as the tradition of "forest monks" has long persisted in southeast Asia (Thailand, Burma, Laos, Cambodia, and Vietnam) in particular.

But the Buddhist perspective is not limited to a kind of primitive nature worship. The Buddha's enlightenment is said to have arisen as a result of his encounter with sickness, old age, and death—that is, after experiencing and thus acknowledging his own and everyone else's eventual encounter with the ultimate unavoidability of something stubbornly real: basic biology. "When I had such power and good fortune," noted the Buddha, recalling his youth as a pampered prince, "yet I thought: when an untaught, ordinary man who is subject to aging, sickness, and death—not safe from them—sees another who is aged, sick or dead, he is shocked, humiliated, and disgusted, for he has forgotten that he himself is no exception."[14] This is a key take-home message of Buddhism: when it comes to those things that really matter (most of which involve the real facts of matter itself), no one is an exception.

Part of the energy behind the rise of Buddhism in the modern West probably comes from what pioneering sociologist Max Weber, borrowing from the German romantic Friedrich Schiller, called the "disenchantment of the world." Buddhism offers, especially to many in the West, an appealing aura of Eastern exoticism combined with the promise of transcending some of the unattractive aspects of modern, high-tech living: notably, its busy focus on money, prestige, and social oppression, often combined with war, environmental destruction, personal frustrations, and anomie. All the better if this perspective, which offers the hope of reenchanting the world, is also one that admits science rather than opposing it.

The next chapter will elaborate the Buddhist awareness of not-self (*anatman*), which, combined with impermanence (*anitya*—see chapter 3), connects seamlessly with what is probably *the* crucial recognition, long known to Buddhists and apparent more recently to biologists as well: the reality of interconnectedness (*pratitya-samutpada*—see chapter 4).

[xxi] To some extent, this phrase is misleading, since everything in the world is "natural," and yet Buddhists and non-Buddhists alike intuitively recognize that some things (trees, flowers, butterflies, lions) are "more natural" than others (garbage dumps, subways, high-rise apartment buildings). Achieving a balance between these constitutes both a challenge and an opportunity.

$\mathbf{A}$n important consideration—which deserves separate treatment that will not be attempted in these pages—is the extent to which "Buddhist modernism," with its emphasis on social, political, and environmental engagement and de-emphasis of such traditional practices as chanting, votive offerings, and other devotional procedures, reflects Buddhism as it was initially promoted by the Buddha and practiced by his early disciples. There are interesting and provocative differences between "originalist" Buddhism, that of the Pali Canon,[xxii] and Buddhism as it has largely been practiced in the West as well as in parts of urban, modernized Asia. These differences are intriguing and worth exploring, especially by historians of religion and comparable scholars. The ensuing debate resembles that between American constitutional originalists and those who maintain that the U.S. Constitution is a living document and should be reinterpreted in the light of modern realities. I vote for the latter.

Every religious tradition struggles with differences between its early scripture and later versions and interpretations. Take Christianity: originally pacifist, it then found itself transmuted—especially via St. Augustine—into a religion that endorsed the notion of the just war. This transition took place largely out of political necessity when the Roman emperor, Constantine, converted to Christianity, whereupon that faith suddenly found itself responsible for defending Rome. Similarly, Buddhism in the twenty-first century finds itself confronting social, economic, demographic, and political realities in addition to the scientific realities of modern biology.

In such cases, the question of authenticity takes a back seat to more pressing considerations, notably whatever happens to work, what seems to be true, what is consistent with human flourishing (as well as the well-being of other living things), and what conveys important and useful guidelines for how to live. There is no reason Buddhism itself shouldn't be susceptible to the same three fundamental Buddhist insights that operate with respect to everything else: not-self, impermanence, and connectedness.[xxiii]

The result is, or at least should be, diminished worry about what constitutes "real" Buddhism. In addition, unlike science, which I believe gets better over time, religions don't necessarily improve with age: sometimes they have gotten worse (e.g.,

---

[xxii] The Pali Canon is acknowledged to contain the earliest records of the Buddha's original teaching.

[xxiii] This tripartite understanding is, to some extent, my own interpretation of Buddhist thought. It should be distinguished from what traditional Buddhism identifies as the "three marks of being," namely impermanence, dissatisfaction (*dukkha*), and not-self.

intolerance on the part of certain radical fundamentalists within Islam, compared with that faith's open-mindedness during the Middle Ages), and sometimes they have gotten better (e.g., the Catholic Church's abandonment of medieval inquisitions and its acceptance—albeit begrudging—of evolution).

When it comes to Buddhism, there is no "perfect," "more real," or "uniquely legitimate" version, although many people—both in Asia and in the West—view at least some aspects of Buddhist modernism with regret. An article in *The New York Times*, dated December 19, 2012, began with this account from the village of Baan Pa Chi: "The monks of this northern Thai village no longer perform one of the defining rituals of Buddhism, the early-morning walk through the community to collect food. Instead, the temple's abbot dials a local restaurant and has takeout delivered." The image of mendicant Buddhist monks, complete with begging bowls, is appealing and picturesque—at least for Westerners, who neither rely on donations for their own meals nor feel obliged to provide food for the monks in question—and its disappearance may or may not be a genuine cultural loss.

On the other hand, many well-informed and admirable Buddhists (including the Vietnamese Zen monk Thich Nhat Hanh and the Dalai Lama) have welcomed what has rightly been termed the "demythologization" and the related "psychologization" of Buddhism in order to make it more compatible with Western culture.[15] Along with this has come a modern emphasis on the "scientific" aspects of Buddhism, an approach that further facilitates the incorporation of Buddhism into American and European intellectual discourse.

At the same time, there is a strain of Western thought that has looked to Buddhism as a potential cultural savior, capable of rescuing us all from the excesses of science, from materialism grown too bold, too encompassing, and yet, at the same time, too dry, inhuman, and downright dangerous. Robert Thurman, for example, has been explicit about hoping that the current Western revival of Buddhism would help to humanize the sciences. It might well do just this, particularly by emphasizing what I call the Buddhist Big Three: *anatman* (not-self), *anitya* (impermanence), and *pratitya-samutpada* ("interconnectedness"). Toward this end I offer the following abbreviated science *sutra*, which this book attempts to flesh out:

Not-self, impermanence, and interconnectedness are built into the very structure of the world, and all living things—including human beings—are no exception.

Buddhism, acting in conjunction with biology, just might help to reenchant our world. No less a materialist than the mathematician, logician, and outspoken atheist Bertrand Russell expressed regret that science had departed from the Greek model of being a "love story between man and nature." As Russell warned, and as many scientists have agreed, there is an ever-present danger that science as an array of strictly materialist explanations might generate a perspective that lacks poetry and substitutes empirical facts and mathematical theorems for underlying meaning. "As physics has developed," wrote Russell,

> it has deprived us step by step of what we thought we knew concerning the intimate nature of the physical world. Color and sound, light and shade, form and texture, belong no longer to that external nature that the Ionians [that is, the Greeks] sought as the bride of their devotion... and the beloved has become a skeleton of rattling bones, cold and dreadful.[16]

Part of the hope lavished upon Buddhism is that it can help animate—more precisely, humanize—this otherwise cold and dreadful skeleton of rattling bones. As we shall see, it is a hope that may well be fulfilled.

2

Not-Self (*Anatman*)

OF ALL BUDDHIST concepts, probably the most difficult for non-Buddhists is *anatman*, "not-self" or—not quite the same, albeit close—"no-self." Bear with it, however, because although counterintuitive, this key notion actually makes perfect sense, especially in the light of organism-environment interpenetration and indeterminacy. As Buddhist masters have known for thousands of years, and "biology masters" have been maintaining for considerably less, the human skin does not separate us from the rest of our environment—it joins us to it. And the more joining there is, the less clear are the boundaries between us and the rest of the world, until it becomes evident that there is no "us" inside distinct from "them" or "it" outside.

The widespread Western conception of self—that of a skin-encapsulated ego—is undergoing scrutiny by physiologists as well as ecologists, geneticists, and evolutionary biologists, and for good reason: insofar as each of us is enmeshed with everything else, the existence of a separate self becomes increasingly untenable. And this recognition is not limited to biologists. Here is Ralph Waldo Emerson, in his essay, *Nature*: "Strictly speaking,...all that is separate from us, all which philosophy distinguishes as the NOT ME, that is, both nature and art, all other men and my own body, must be ranked under this name, NATURE." Later, by the time of his renowned Harvard lecture "The American Scholar," Emerson's "not me" had Buddhistically morphed into the "other me": "The world—this shadow of the soul, or other me, lies wide around. Its attractions are the keys which unlock my thoughts and make me acquainted with myself."

What about those thoughts, and that self?

René Descartes was a brilliant man, the originator of Cartesian geometry, and a true pioneer not only in mathematics and philosophy but also in the dangerous (at the time) discipline of human anatomy and dissection. He is best known these days as the fellow who proclaimed *cogito ergo sum*—"I think, therefore I am"—as he sought something firm and indisputable upon which to build his worldview. Ambrose Bierce, incidentally, modified Descartes's dictum to *cogito cogito ergo cogito sum*—"I think I think, therefore I think I am"—arguing that this was as close to certainty as philosophy was likely to get!

In any event, and although it might seem un-Buddhistically critical, it must be noted that when it came to his own thinking about thinking, Descartes blew it. He thought he was starting with something indisputable, but in fact, Descartes's basic assumption has for centuries misled philosophers and biologists, though not Buddhists. Now neuroscientists are catching up. Descartes was the intellectual progenitor of the dualism that bears his name, the presumption that mind and body—more precisely, mind and brain—are separate. According to the neurobiologist Antonio Damasio, "Descartes' Error" was his separation of mind from brain.[1] I grant that this indeed was a mistake, but Descartes's bigger error was even more fundamental: separating "I" from the rest of the world.

"A human being is part of a whole," wrote Albert Einstein,

> called by us the "Universe"—a part limited in time and space. He experiences himself, his thoughts, and feelings, as something separated from the rest—a kind of optical delusion of his consciousness. This delusion is a kind of prison for us, restricting us to our personal desires and to affection for a few persons nearest us. Our task must be to free ourselves from this prison by widening our circles of compassion to embrace all living creatures and the whole of nature in its beauty.[2]

There is a marvelous scene in the movie *Men in Black* in which a body is brought into a morgue. Upon dissection, it turns out that it wasn't really a person after all, but a realistic robot controlled by a little green alien who resided—where else?—in its head, from which he manipulated the levers of movement. This is Cartesian dualism with a vengeance, depicting a literal dichotomy between mind—the "me" represented by the little green homunculus—and that lumbering conglomeration of muscle, bone, blood, and nerves that we call our "self." (Of course, to be consistent, that internal creature must also be a kind of robot, with yet another little green homunculus inside *its* head, and so on, a retinue of Russian nesting *Matryoshka* dolls, ad infinitum. But that's another story.)

Despite the fact that everything we know about neurobiology makes it overwhelmingly, undeniably, unquestionably clear that dualism is nonsense—that our minds are nothing but the result of mechanical processes occurring within our brains—dualism remains stubbornly appealing, almost a default setting when it comes to conceptualizing our own mental processes. Thus, Francis Crick—codiscoverer of the molecular structure of DNA and one of the greatest scientists of the twentieth century—felt it altogether appropriate to title his book on the biology of consciousness *The Astonishing Hypothesis* and to begin it as follows:

> The Astonishing Hypothesis is that "You," your joys and your sorrows, your memories and your ambitions, your sense of personal identity and free will, are in fact no more than the behavior of a vast assembly of nerve cells and their associated molecules. As Lewis Carroll's Alice might have phrased it: "You're nothing but a pack of neurons."[3]

Francis Crick spent the last years of his life attempting to elucidate that hypothesis, to trace the details whereby neurons, electrochemical cascades of charged ions and proteins, peptides, and various neurotransmitters crossing semipermeable membranes somehow produce all those ineffable internal sensations that make up consciousness. Like Einstein, whose last decades were absorbed in attempting to come up with a "grand theory of everything" that reconciled quantum and relativistic mechanics, and failing, Crick also fell short. But unlike Einstein's effort—which is not guaranteed success (there just might not be such a reconciliation)—Crick's quest is, at least in theory, guaranteed not to be quixotic: it really would be astonishing if consciousness did *not* reside, somehow, in the material world of neurobiology.

The Dalai Lama himself has long had a genuine scientific interest in mind-brain correlations, such that he was the invited plenary speaker at the huge Society for Neuroscience annual meeting in November 2005. He eventually spoke on the neuroscience of meditation, but his very invitation caused an uproar. There was a protest petition that garnered about a thousand signatures, mostly from scientists worried about religion invading science and thereby degrading it. In any event, the Dalai Lama touched on something that has also received a great deal of attention, probably much more than it deserves: namely, the question of whether meditation actually causes bona fide changes in brain function among those who engage in it. Evidently it does.

Indeed, the scientific world—not just that of laypersons—was abuzz when a group at the Waisman Laboratory for Brain Imaging and Behavior at the University of Wisconsin–Madison, led by the professor of psychology Richard J. Davidson, reported in the *Proceedings of the National Academy of Sciences* (probably the most prestigious scientific publication in the United States), that Tibetan Buddhists who had practiced serious meditation for many years had brain-wave patterns that differed consistently from those of a nonmeditating control group.[4] But why should anyone have been surprised? The Buddhist meditators had engaged in years of prolonged and, indeed, arduous mental training that involved from ten thousand to fifty thousand hours of serious meditation. Compared to "healthy student volunteers," they exhibited "high-amplitude gamma synchrony." Wouldn't it be even more surprising if such experiences didn't generate some sort of discernible effect in their brains?

If you examine the brain functioning of people who have been watching seven hours of television per day, I daresay you would find changes in their cerebral functioning, too. Ditto for anyone reading this book, or yearning to scratch an itch. The point is not to disrespect the meditation-neurobiology research results, but to note the degree to which even educated, scientifically sophisticated individuals remain incredulous at the revelation that the mind and brain are connected. Somehow, the very fact that meditation generates reportable brain changes was—and still is— widely seen as making this ancient Buddhist practice more legitimate.

At the same time, and even as we can announce with certainty that the mind resides wholly and entirely in the brain and that it is wholly and entirely material— which is to say, anatomical, chemical, electrical, metabolic, and so forth—everything that neuroscientists are learning and have learned confirms that there is no single, simple "I" inside anyone's head. No little green alien, no homunculus, no entity to constitute Freud's "observing ego." Nobody home. You can parse the brain, probe it, and stimulate it to your heart's content, in the process stimulating—or eliminating—subjective experiences, but just as Renaissance theologians were distressed not to find God when they looked through Galileo's telescope, no one will find himself or herself, or anyone else, inside a brain.

The English word "person" comes from the Latin *persona*, for mask. What is behind the mask? Who is wearing it? For Buddhists and, increasingly, neurobiologists, the answer is, "No one." Although this might seem a logical impossibility, bear with me; the outcome is less peculiar than it appears.

Every human being possesses something like a hundred billion neurons, each one connected to hundreds—sometimes thousands—of others in astoundingly complex ways, resulting in patterns and interactions that come about as close to infinite as any finite numbers are likely to get. These neurons are whispering almost constantly to each other. As a result, our perceptions have a kind of continuity; but as Buddhists have long maintained and neurobiologists can now confirm, such continuity is a fiction. Think, for example, about watching a movie: people in the audience are actually seeing a parade of still images, which, if they follow upon each other rapidly enough (roughly twenty-four frames per second) give the illusion of a smooth, uninterrupted, continuous flow.

Harold Fromm gives an account of another optical illusion, resulting once again in a false impression of continuity:

> In the case of our eyes, 100 million rods and 7 million cones in our retinas—the receptors of light from the scenes we behold—send electrochemical impulses to the brain via neurons. Since our range of clear sight consists of a very small area directly in front of us, we are constantly refocusing our eyes and moving our head at the rate of several "saccades" (or eye movements) per second. This means that the smooth-seeming panorama that we view is completely redrawn several times a second, since every rod and cone receives a different light particle with each refocus. Just as we don't hear the 44,000 interruptions per second between the samples of music coded on a compact disk, don't see the individual frames of a movie film or the many redrawings per second of our TV and computer monitors, we are completely unaware of the pointillistic nature of our vision.[5]

Of what, then, are we aware? Or to put it differently, what *are* we—mentally—aside from the awareness that occupies our selves at any given moment? The philosopher David Hume pondered this question some centuries ago and concluded: we are nothing at all. "There are some philosophers," wrote Hume in *A Treatise on Human Nature,*

> who imagine we are every moment intimately conscious of what we call our SELF; that we feel its existence and its continuance in existence; and are certain...both of its perfect identity and simplicity. The strongest sensation, the most violent passion, say they, instead of distracting us from this view, only fix it the more intensely....But there is no impression constant and invariable. Pain and pleasure, grief and joy, passions and sensations succeed each other, and never all exist at the same time.

In short, there is no self capable of sitting back and observing these sensations, independent of the sensations themselves.

This is dangerous ground indeed, at least for anyone wedded to the idea of a distinct, enduring self that occupies our head and is in charge, director of every individual's personal Central Intelligence Agency. Among the many problems that arise from the lack of a clearly identified self, a big one concerns that perennial problem, free will. After all, if we are merely participants in an array of instantaneous perceptions, where does volition come from? Where is the "I" who makes up his or her mind about, well, anything? Once we have banished the little green man or woman, how are we to respond if we ever meet that other fictional little green guy, the one who arrives in a flying saucer and demands, "Take me to your leader"? Some recent findings in neurobiology make a coherent response even more problematic.

To arrive at one such finding, Benjamin Libet conducted a surprisingly simple study of conscious volition, the straightforward idea that "you" make up "your mind," before doing something. How often do we say, "Wait a minute, I'm making up my mind"? Eventually—whether after a minute, an instant, a month, or more—we do so. We decide on something, which generally means that we stand by while this decision progresses to our consciousness. From where, to where? And who and what is the pilgrim that makes this progress? In Libet's research, the test subject was asked to sit comfortably and then, whenever he or she decided, to move a hand. No constraints, no particular signal to follow. Just move your hand, the subjects were told, whenever you feel like it. It had already been found that about eight-tenths of a second before anyone actually performs a movement—any movement—his or her brain engages in a characteristic pattern of wave activity known as a Readiness Potential (RP). Libet's subjects were wired to EEG machines so as to record the onset of each RP, but he added a twist: each subject was to look at a rapidly moving clock face, designed to make it easy for him or her to note the precise time at any instant. The task, then, wasn't just to wiggle your hand whenever you felt like it, but also to note the exact moment when the decision was consciously made.

It is no great surprise that there was a delay between each decision to wiggle and the actual movement; it takes time, after all, for a nerve impulse to travel from brain to hand and for the muscles to respond accordingly (about one-third of a second). The strange thing was this: the "conscious decision," as reported by Libet's subjects, always took place another one-third to one-half second *after* the associated RP. In other words, undeniable electrical evidence showed that subjects were actually "ready" for the action—in some sense, had already "decided" to perform it—before they knew that they had.[6] And this is not a murky, diffuse, unconscious yearning or a bizarre dream emanating from a dark, dank Freudian pit of mystery, but a simple, intentional, and presumably conscious movement of one's hand. Each of Libet's subjects had no doubt that he or she was in control of these movements, but it is difficult not to conclude that in fact, rather than a conscious, intentional controller, each subject was an observer, not unlike the experimenter and others in the lab who were observing these same subjects.

As with volitional movement, we might well conclude, so with consciousness itself. When closely examined, what seems coherent, consistent, and under our control is found, instead, to be strangely lacking in underlying substance—at least in part because there isn't anyone in the pilot's seat.

Sir Charles Scott Sherrington, who, in 1932 received the Nobel Prize in Physiology or Medicine, studied the reflex circuitry between nerves and muscles and coined the term "synapse." This same biologist once described the brain as "an enchanted loom where millions of flashing shuttles weave a dissolving pattern."[7] The shuttles flash, and the loom weaves a pattern that constantly dissolves and, accordingly, can never be grabbed, any more than we can freeze and thus perceive the individual frames that make up a motion picture. It may win an Academy Award, but the movie doesn't really exist. Not only that, but there isn't anyone watching it. By the same token, "there is no such thing," wrote Ludwig Wittgenstein, "as the subject that thinks or entertains ideas."[8] In short, there are thoughts, but no thinker.[i]

There is also this parallel, from the *Visuddhimagga* (The Path to Purity)—a renowned Theravada Buddhist commentary written by the scholar Buddhaghosa in approximately 430 CE in what is today Sri Lanka: "There is suffering, but none who suffers; Doing exists although there is no doer. Extinction is, but no extinguished person; Although there is a path, there is no goer."

Although there are brain waves, a veritable hurricane of neural activity within every sentient brain, there is no homunculus. Although there are thoughts and even actions, it isn't even clear that there is anyone responsible. No one home. No self. *Anatman.*

Before we go on, this might be a good place to introduce, briefly, the question of translation, since along with *anatman*, we'll be encountering a number of other Sanskrit words, notably *anitya* and *pratitya-samutpada* to refer, in turn, to "not-self," "impermanence," and "interconnectedness." Any act of translation necessarily becomes one of transformation as well, something that in the process runs the risk of misrepresentation. In his insightful book *The Making of Buddhist Modernism,*[9] David McMahan confronts two examples of how common Buddhist words (*moksa*

---

[i] Nonetheless, it can certainly be argued that our sense of self is a necessary (and natural) construct that gives coherence to our experience, much as optical saccades enable us to create continuous visual experience.

and *bodhi*) can be unintentionally twisted and thus somewhat misunderstood when translated:

> We might, for example, translate the Sanskrit term *moksa* as "freedom." In Buddhism, this means liberation from rebirth in samsara as an embodied being, as well as liberation from destructive mental states (*ikesas*), craving, hatred, and delusion and from the suffering (*duhkha*) they produce. When, however, *moksa* is translated as "freedom"—a perfectly justified translation, by the way—it cannot help but pick up the tremendous cultural resonances this word has in modern European languages and cultures. It inevitably rings the notes of individual freedom, creative freedom, freedom of choice, freedom from oppression, freedom of thought, freedom of speech, freedom from neuroses, free to be me—let freedom ring, indeed. It is virtually impossible to hear this word without a dense network of meanings lighting up, meanings deeply implicated in the history, philosophies, ideologies, and everyday assumptions of the modern West.
>
> The translation of one of Buddhism's central terms, *bodhi*, provides another example. It literally means "awakening" and describes the Buddha's highest attainment under the bodhi tree. The most common English translation, "enlightenment," invokes, however, a complex of meanings tied to the ideas, values, and sensibilities of the European Enlightenment: reason, empirical observation, suspicion of authority, freedom of thought, and so on. Early translators, moreover, consciously forged this link. Buddhist studies pioneer Thomas W. Rhys Davids (1843–1922) first translated bodhi as "Enlightenment" and explicitly compared the Buddha with the philosophers of the European Enlightenment.[10]

As McMahan recognizes, this does not mean that we need to find other ways to translate *moksa* and *bodhi*; rather, it simply emphasizes that translation inevitably involves changing the original, sometimes enlarging it, sometimes deforming it but always modifying it. Like individual bodies—not bodhis!—and their constituent selves, words, too, are subject to *anatman* and also to *anitya* and *pratitya-samutpada*.

Here is the title of a very successful self-help book, written several decades ago: *Our Bodies, Our Selves*. It announced, somewhat defiantly, a widespread notion, one that is so obviously true that it is rarely questioned or even acknowledged. Whatever is true—or false—about that inner, conscious entity we call our self, nothing seems more obvious than the fact that it is somehow contained within something else: a body. If the body isn't the embodiment of our self, then what is? And, of only slightly lesser moment: if our body is not our self, then what is it? Can there be any legitimacy to the Buddhist assertion of "not-self" applied to that bastion of the singular "us," the identified "me," our own bodies?

Who, in short, would be so foolish as to take issue with the existence of the i-bod? Buddhists, for some. And biologists, for others.

For a statement of *anatman* from today's foremost modern exponent of Buddhism this side of the Dalai Lama, consider the following thought-provoking and oft-cited observation by Thich Nhat Hanh, a beloved Vietnamese monk in the Zen tradition, whose antiwar activities earned him the enmity of both the North Vietnamese and the US governments, and who was also nominated by Martin Luther King, Jr. for a Nobel peace prize. Thich Nhat Hanh asks us to look deeply into the physical realities of the world, pointing out:

If you are a poet, you will see clearly that there is a cloud floating in this sheet of paper. Without a cloud, there will be no rain; without rain, the trees cannot grow; and without trees, we cannot make paper. The cloud is essential for the paper to exist. If the cloud is not here, the sheet of paper cannot be here either. So we can say that the cloud and the paper inter-are. "Interbeing" is a word that is not in the dictionary yet, but if we combine the prefix "inter-" with the verb "to be," we have a new verb, inter-be. Thus, the cloud and the sheet of paper inter-are. If we look into this sheet of paper even more deeply, we can see the sunshine in it. If the sunshine is not there, the forest cannot grow. In fact, nothing can grow. Even we cannot grow without sunshine. And so, we know that the sunshine is also in this sheet of paper. The paper and the sunshine inter-are. And if we continue to look, we can see the logger who cut the tree and brought it to the mill to be transformed into paper. And we see the wheat. We know the logger cannot exist without his daily bread, and therefore the wheat that became his bread is also in this sheet of paper. And the logger's father and mother are in it too. When we look in this way, we see that without all of these things, this sheet of paper cannot exist. Looking even more deeply, we can see we are in it too. This is not difficult to see, because when we look at a sheet of paper, the sheet of paper is part of our perception. Your mind is in here and mine is also. So we can say that everything is in here with this sheet of paper.

You cannot point out one thing that is not here—time, space, the earth, the rain, the minerals in the soil, the sunshine, the cloud, the river, the heat. Everything co-exists with this sheet of paper. That is why I think the word inter-be should be in the dictionary. "To be" is to inter-be. You cannot just be by yourself alone. You have to inter-be with every other thing. This sheet of paper is, because everything else is. Suppose we try to return one of the elements to its source. Suppose we return the sunshine to the sun. Do you think that this sheet of paper will be possible? No, without sunshine nothing can be. And if we return the logger to his mother, then we have no sheet of paper either. The fact is that this sheet of paper is made up only of "non-paper elements." And if we return these non-paper elements to their sources, then there can be no paper at all. Without "non-paper elements," like mind, logger, sunshine and so on, there will be no paper. As thin as this sheet of paper is, it contains everything in the universe in it.[11]

Now, turn your attention from the cloud in that sheet of paper to the cloud in your own skin, the carrot in your kidneys, the dirt in your pancreas, the bit of fossil fuel in your biceps, the components of nitrogenous fertilizer molecules in your liver. Every part of "me" is composed of "not-me" elements, and ditto for "you." In fact, there is nothing in me or you that is anything *other* than not-me or not-you elements. This may well seem downright weird and altogether abstract.

And yet, a Buddhist perspective on the self is nothing if not practical. Buddhists aren't blind, or stupid, or unaware that everyone has a body. They are not so foolish as to deny the existence of bodies as we all know them, or the convenience and even the legitimacy of identifying a person with his or her body. To say that the self is not a distinct essence or enduring entity (i.e., to profess agreement with the Buddhist conception of *anatman*), is not the same as saying that the self doesn't exist at all.

I don't know it for a fact, but I imagine that the Dalai Lama didn't simply dispatch a disembodied, ethereal essence, which wafted wispily from his exile in Dharmasala, India, to Washington, D.C., to attend the neuroscience conference discussed here. Rather, I'll bet he had a ticket, made out in his name, guaranteeing him a seat in which he placed his altogether corporeal body. Moreover, a twenty-first-century Buddha, were he a US citizen, would doubtless have his own Social Security number, not to mention a driver's license, adorned with a photograph—*his* photograph, a pictorial record of his bodily substance.

So what does it mean to proclaim the reality of *anatman*? And what of those selves—Buddhist or otherwise—who make this proclamation?

Here is a useful way of thinking about it, from Stephen Batchelor: "The point is not to dwell on the absence or emptiness" but to "encounter the phenomenal world in all its vitality and immediacy once such a conception of self begins to fade." Batchelor illustrates this with a persuasive example from his own life:

When my wife and I bought our house in France, a large wooden shed stood in the garden just behind the house itself. The shed cut out the light and blocked the view. Moreover, it had become overgrown with honeysuckle and ivy, so that its size increased year by year, darkening the shadow it cast and increasing the rank humidity in the passage between it and the house.... Finally, we got rid of the shed.... In the course of an afternoon, something that had been such a gloomy presence for so long was not there anymore. For the next few days, I would stand where it had once stood and consciously delight in its absence. The dark, dank passage it created had vanished. House and garden were transformed. Light poured into the downstairs windows, and hitherto unknown vistas of the garden and surrounding countryside opened up.

After a few days, the rapturous experience of "no shed" faded. I forgot about the shed that had once been there and was no longer struck by its absence. I simply attended to the garden and house as they now were. For Dharmakirti [a seventh-century Indian Buddhist monk and scholar], the experience of "emptiness" or "not-self" is like this. To realize the absence of a permanent, partless, autonomous ego enables hitherto unknown vistas in one's life to open up. The dark, opaque perspective of self-centeredness gives way to a more luminous and sensitized awareness of the shifting, contingent processes of body and mind. Once one gets used to this, one ceases to notice the absence of such a self. It is replaced by another way of living in this world with others, which, after a while, becomes entirely unremarkable. To keep insisting on "emptiness" as something sacred and special would be like erecting a shrine in the garden to the absence of the shed instead of getting on with the gardening.[12]

Okay, readers, let's get on with the gardening!

In traditional Buddhist thought, human "selfhood" consists of Five Aggregates: 1. its material or bodily form; 2. sensations, which are pleasant, unpleasant, or neutral; 3. perception, by which we identify objects as well as have ideas and thoughts; 4. mental formations, which consist of the various attitudes, character traits, and dispositions that lead to actions and thus are important and subject to free will; and 5. consciousness, which is essentially sensory as well as mental awareness. But in no case is there a single or permanent self. Using the metaphor of self as a chariot, a Buddhist nun named Vajira explained: "Just as, with an assemblage of parts, the word 'chariot' is used, so, when the aggregates are present, there's the convention of 'a being.' "[13]

For biologists, however, the fiction of individual beings has been useful. How else could researchers in animal behavior describe what something that certainly looks like a particular wolf, for example, is doing when it stalks a moose, emits a nocturnal howl, or attempts to copulate with another conjunction of molecules that is identifiable as another wolf? But many concepts are widespread, seemingly obvious, often convenient—and nonetheless wrong. Take the obvious "fact" that the earth is flat, or that the sun rises and sets, or even, that matter is solid rather than—as modern physics knows for certain—shot through with empty space. Body-centeredness is in this category: useful, widespread, seemingly obvious—yet wrong. In fact, one of the most important advances in modern evolutionary biology has come from the recognition that bodies are not nearly what they had been cracked up to be. Not only that, but biologists found that when they recognized the transient, contingent nature of bodies, the result was nothing less than a revolution in their understanding of evolution.

At the core of this revolutionary biology is a new perspective on the relationship of bodies and genes.[14] Bodies have been demoted; genes, promoted. Not that bodies aren't important. Look around: it's bodies that we see occupying the landscape, whether pumpkins, poodles, or people. Bodies, bodies everywhere, bodies breathing, reproducing, fighting, sharing, arguing, running about, lying down, building buildings, writing novels, voting, emoting, studying Buddhism, making war, and making love. Every person is carefully tucked into a body, occupying—or so it is non-Buddhistically assumed—neither more nor less than the exact boundaries of his or her own private integument.

Or, as an old song by the parodist Alan Sherman proclaimed, "You Gotta Have Skin"—to the tune of an even older song, "You Gotta Have Heart":

You gotta have skin
All you really need is skin.
Skin's the thing that if you've got it outside,
It helps keep your insides in.
You gotta have skin.

For all its silly humor, this little ditty does a good job of encapsulating a Western, distinctly non-Buddhist perspective (even as "encapsulating" is itself non-Buddhist in its focus on separation).

Biologists often make a useful distinction between "skin-in" and "skin-out" aspects of their science. The former refers to the study of what transpires within a creature—its anatomy, physiology, cell biology, and so forth—while the latter takes living things as more or less independent entities and concerns itself with how whole organisms interact with each other and with their environment: it is the realm of ecology, ethology (animal behavior), evolution, and so on. This distinction between skin-in and skin-out biology is widespread in the discipline; it is also commonsensical and obvious, but deceptive. Consider, for example, that the best view in Warsaw is from the top of the Palace of Culture and Science, a monument to Stalinesque architecture at its worst. Why is the view so good? Because it's the only place in the city from which you cannot see the Palace of Culture and Science! Similarly, each of us looks out at the rest of the world from the private confines of our cozy little home and castle: our bodies, ourselves. Skin-inners looking at the skin-outness of other, similar creatures. The view is good, but not completely accurate.

It is very difficult for people to consider themselves anything other than their bodies, taken as a whole. You may look, with detachment, at a sample of "your" red blood cells, wriggling on a microscope slide, at a dental X-ray, a chromosome enlargement, or an MRI of your intestines, all the while knowing that in some sense, it isn't truly you.

From the perspective of evolution, who or what constitutes that individual you call yourself? And what does it mean to acknowledge that a lengthy process of natural selection has produced it? Taken at face value, insofar as every living thing is the product of evolution, each individual "me" has somehow been squeezed and molded, pruned and pounded into shape by natural selection operating via natural selection's impact—not on ourselves, but on our ancestors.

Correct, but only up to a point. Everyone is indeed the product of evolution. Biology's revolution has shown, however, that the face that stares back from the mirror (just like the one who looks out at the world from that structural bastion we call a body) is only indirectly evolution's handiwork. Rather, all bodies are the immediate product of genes, and it is they that are evolution's legacy to every person, made corporeal in all those bodies that people today's world. Your face, your body, your physical and biological self: all are the outer manifestations of your genes after they have interacted with the food, the oxygen, the learning, the exercise, and the sum total of all the other experiences that eventually produced you. It is human genes— and hummingbird genes, and hyacinth genes, and so on—that have been picked and pruned, some doing well and others poorly, tumbling over each other, some leaping ahead and others dying out, mutating on occasion or remaining largely unchanged, flowing along as best they can in a continuously meandering stream of DNA for millions of years, beginning with the origin of life and with no end in sight.

"We are but whirlpools in a river of ever-flowing water," wrote Norbert Wiener, founder of cybernetics and one of the great mathematicians of the twentieth century. Wiener went on: "We are not stuff that abides, but patterns that perpetuate themselves."[15] These patterns clearly exist, at any given moment, in a form we recognize as a body, but each individual is more like a temporary eddy in that living, surging river.

You are only as old as the time that has passed since your birth. In a sense, it wasn't evolution that put you together so much as the laws of chance, which combined half of your mother's genes with half of your father's, then shook you gently with a hefty dose of experience and environment, finally marinating the concoction in tincture of time. A famous Zen koan—a riddle intended to shake seeker-monks free of their excessive dependence on linear thought processes—asks the novitiate to describe his face before he was born. A good answer these days would have something to do with DNA, but it would also have to incorporate ebb and flow, process and pattern, contingency and probability, time and a flowing river.[ii]

The genes that carry the information needed for each cell and protein in your body were established in the eons that passed before you were born. Like playing cards, they have the potential to be reshuffled and reused in different combinations, over and over again in a poker game that lasts not merely all night, but potentially for all time. You, by contrast, won't be around nearly so long.

Each individual is analogous to a hand of bridge, pinochle, gin rummy, hearts, or poker (take your pick—though there are many more potential people than potential arrangements of playing cards). Over an evening of card playing, such "hands" are evanescent. It is the game—of life—that persists, and the cards—genes—that make it possible. At the same time, there are many ways of playing out the hand you are dealt. Your face, its scars and lines and wrinkles, straight or crooked teeth, glasses or no glasses, no less than your sentiments, beliefs, memories, and dreams, are unique to you and your personal developmental history, just as a hand of cards is unique to a given occasion. Sometimes, in the long road from potential to actual (biologists would say, from genotype to phenotype—more on this later), your particular genetic legacy is hidden: although you may have been born with genes for crooked teeth, if you had orthodonture, your smile today reflects not only those genes but also a good dose of corrective—not to say expensive—experience.

But no matter how much money, time, or effort is lavished on them, bodies don't have much of a future, and this is where *anatman* comes in, yet again. In the scheme of things, bodies are as ephemeral as a spring day, a flower's petal, a gust of wind. What does persist? Not bodies, but genes. Bodies go the way of all flesh: ashes to ashes and dust to dust, molecule to molecule and atom to atom. Bodies, schmoddies. They are temporary, cobbled together from various recycled parts scavenged from the cosmic junk heap and only temporarily on loan to you. Genes, on the other hand, are potentially immortal. (We'll look at the phenomenon of impermanence— in BuddhaSpeak, *anitya*—in the next chapter.)

In his poem "Heredity," Thomas Hardy had a premonition of modern evolutionary biology and the endurance of genes, which is to say, of Buddhist biology:

I am the family face
Flesh perishes, I live on

[ii] This seems to be an interesting example of (molecular) biology catching up with Buddhist thought, such that we might well need a new koan insofar as the goal is to promote nonlinear thought.

Projecting trait and trace
Through time to times anon,
And leaping from place to place
Over oblivion.
The years-heired feature that can
In curve and voice and eye
Despise the human span
Of durance—that is I;
The eternal thing in man,
That heeds no call to die.

When people refer to evolution with the verbal shorthand "survival of the fittest,"[iii] only rarely do they understand just what they are talking about. The fittest, as today's biologists use the term, are those who are most successful in projecting copies of their genes into the future. More precisely yet, and more in tune with revolutionary biology: the fittest genes are those that succeed in promoting their own success. Fitness is what genes strive for; bodies are how they do it. (I'll have something to say later about striving, as do Buddhists.) Aside from that, bodies don't count, don't matter, and although they assuredly exist, their mere existence in the world of matter—even their longevity—is not the point.

Indeed, most individuals of most species are actually rather short-lived. In the billions of years that evolution has had at its disposal, natural selection has perfected all sorts of highly sophisticated, complex adaptations, absolute marvels of highly functional design: mammals that fly and that use sonar to hunt moths on the wing, caterpillars that look like snakes and vice versa, bacteria that can prosper in super-heated underwater deep-sea vents, eagles that can make out the face of a dime while hovering a hundred feet in the air.

Amid all these accomplishments, if there had been a significant evolutionary advantage for individual organisms to live for hundreds of years, it seems very likely that this could have been done, too. After all, in some cases it has. Redwood trees and bristlecone pines survive for thousands of years. And yet, these are notable exceptions. The answer—first pointed out by George C. Williams, one of the founders of revolutionary biology—is that mere survival of the individual is not very high on the evolutionary agenda; in fact, the overwhelming majority of living things have a lifespan of less than a year.

[iii] This phrase, incidentally, was coined by the social philosopher Herbert Spencer, and not by Darwin. Its use has bedeviled biologists and confused many laypersons ever since, especially because it is often mistaken to refer to physical fitness, whereas it only makes biological sense as a measure of success in projecting genes into the future.

It all makes sense, once we recognize that the individual exists for the benefit of the genes, not the other way around. Hence, a chicken really is an egg's way of making more eggs; in fact, from a strictly biological perspective, it is nothing but. This is a way of understanding nature, not a value judgment or political statement on the importance of individual lives, human or otherwise.[iv] By analogy, we can say that the planet Earth is only a tiny ball of water, mud, and metal in a vast universe. But it is our Earth, our home, and very important to us, just as each individual, body, mind, and spirit is likely to be critically important to each human being.

Sigmund Freud, of all people, prefigured a similar perspective, emphasizing not surprisingly sex. In *On Narcissism*, Freud observed that

> The individual himself regards sexuality as one of his own ends; whereas from another point of view he is an appendage to his germ-plasm, at whose disposal he puts his energies in return for a bonus of pleasure. He is the mortal vehicle of a (possibly) immortal substance—like the inheritor of an entailed property, who is only the temporary holder of an estate which survives him.[16] [Note to reader: Please forgive Dr. Freud's sexist pronouns. It is probably fair to attribute this—or at least, most of it—simply to the prevailing linguistic sexism of his day.]

L et's go back to Thich Nhat Hanh, he of the cloud-in-the-sheet-of-paper and many other beautiful and provocative images, who writes that we need to "awaken from the illusion of our separateness." Please note, by the way, that this doesn't necessarily require a scary, acid-trippy, or semipsychotic loss of all boundaries, as exemplified in a lovely but disturbing story from the Vedic Hindu tradition: "A grain of salt wanted to ascertain the saltiness of the ocean, so she threw herself into the sea, thereby gaining a fully accurate, perfect perception!" By contrast, the Buddha's individuality wasn't annihilated just because he lost (or rather, went beyond) the illusion of a distinct, separate self. Far from it: the Buddha continued to teach for another forty-five years, had numerous encounters with all sorts of people, and expressed himself in unique and memorable ways—sometimes poetically, sometimes humorously, sometimes (rarely) even revealing an impatient side.

---

iv A similar argument, I believe but cannot confirm, was made by Martin Heidegger to the effect that "we are something that the world is doing."

Some of his impatience was evoked by the ancient Brahmanical texts known as the *Upanishads*, which predated Buddhism by a thousand years and more, and against which the then-emerging tradition of Buddhism had to define itself—not unlike Christianity struggling to distinguish itself from its ancestral Judaism. The *Upanishads* promote a very different conception of the self, or *atman*, as an unchanging substance (not altogether different from the Christian sense of a soul) that is eternal and uniquely associated with each individual. For Hindus, *atman* is not only indestructible but also beyond description as well as beyond suffering: "The self is not this and not that. Ungraspable it is certainly not grasped; indestructible it is certainly not destroyed, without clinging it is certainly not clung to; unbound it comes to no harm, it does not suffer."[17] This *atman*, in classical Hindu thought, is connected to the underlying ground substance of all reality, namely *brahman*. For the *Upanishads*, it isn't too much of a stretch to suggest that *atman = brahman*. Buddhists, as we have seen, view things differently, as reflected in the alternative designation, *anatman* or "not-self."

Once again, it must be emphasized that not-self does not mean nonexistence. Buddhists are not fools. "If we arrive at the knowledge that the self at which we grasp is empty," writes the Dalai Lama,

> we may imagine this means that we as individuals with personal identities do not exist. But of course this is not the case—our own personal experiences demonstrate that as subjects and agents of our own lives, we certainly exist. So how, then, do we understand the content of this insight into absence of self? What follows from this insight? We must be very clear that *only the self that is being grasped as intrinsically real* needs to be negated. The self as a conventional phenomenon is not rejected. This is a crucial aspect of the Buddha's teachings on emptiness. Without understanding this distinction, one cannot fully understand the meaning of no-self.[18] [italics in original]

To this should be added that emptiness—*anatman*—itself is also devoid of intrinsic existence. And so are Buddhas.

Even genes, which modern biologists increasingly see as the primary units of selection and thus, the entities whose relative abundance and rarity, whose successes and failures drive the process of natural selection—even these seemingly distinct blueprints for the construction of organisms are themselves notably squishy (conceptually, if not structurally) in that they don't translate directly into the various aspects of a living thing, its phenotype, that we actually perceive when we consider an individual plant, bacteria, animal, or human being.

For example, even the evolutionary theorist Richard Dawkins—who has done more than any scientist to convince the lay public, as well as his fellow biologists, of the importance of taking a gene's-eye perspective on the evolutionary process—has made it clear that

> the phenotypic "effect" of a gene is a concept that has meaning only if the context of environmental influences is specified, environment being understood to include all the other genes in the genome. A gene "for" A in environment X may well turn out to be a gene for B in environment Y. It is simply meaningless to speak of an absolute, context-free, phenotypic effect of a given gene.[19]

It is little appreciated by the lay public that when geneticists talk loosely about a gene "for" something, they are actually using shorthand for a more complex and crucial concept: namely, they are really talking about the difference between at least two phenotypes, each produced by a different allele. Hence, there is no phenotype encapsulated within genes; rather, there is the potential for creating phenotypes of differing kinds. Genes are not separate, biologically meaningful entities like beads on a string, even though when it comes to the mathematical calculations of population geneticists and theorists, they are often treated as though they are. This is a benevolent deception, engaged in simply to achieve what mathematicians (who are increasingly involved in this kind of research) refer to as "tractability."

Actually, we deal with lots of other constructs, which, like the self, aren't real, yet are treated as though they are. Consider money, for example: it isn't of any intrinsic value, yet we take it otherwise. It's a carefully orchestrated group mythology, something that has meaning only insofar as we agree to grant it validity. For the anthropologist Weston LaBarre, money is "an institutional dream that everyone is having at once."[20] A dollar bill, for example, isn't really of much inherent value, except maybe to help light a fire, but when others agree to accord it value (e.g., $1), it acquires meaning beyond that to which it is otherwise entitled. Something very similar may well have happened—especially in the non-Buddhist West—with respect to the idea that each of us is a separate self.

At the same time, Western thought has not been monochromatic when it comes to selfhood and individuality. Friedrich Nietzsche called for *Selbstüberwindung*, or "self-overcoming," just as the Buddhist *Dhammapada* teaches that "although he may conquer a thousand times a thousand men in battle, yet he indeed is the noblest victor who would conquer himself," which in turn is paralleled by Islam's distinction between the "lesser" and the "greater" jihad, with the former defined by war against Islam's external enemies and the latter by the more difficult struggle with one's self.

And consider this verse from Goethe's poem *The Mysteries*:

All force strives forward to work far and wide
To live and grow and ever to expand;

Yet we are checked and thwarted on each side
By this world's flux and swept along like sand;
In this internal storm and outward tide
We hear a promise, hard to understand:
From the compulsion that all creatures binds,
Who overcomes himself, his freedom finds.[21]

When discussing the makeup of what passes for an individual, Buddhist thought emphasizes the existence of the Five Aggregates (*skandhas*), which constitute a complete description of our subjective "selves." In order of increasing sophistication and complexity, these *skandhas* begin with (1) the basic physical body. Then comes (2) direct mental experience or feelings which derive from that body: pleasant, unpleasant, and neutral sensations. These inputs are constantly (3) sorted, classified, and identified, enabling us to recognize what is being perceived, which in turn gives rise to (4) various "mental formations" and desires, perhaps sexual attraction to certain individuals, hunger for particular foods, exaltation at an especially satisfying thought and/or aversion from an unpleasant one, and so forth. At the apex of *skandhas* we encounter (5) consciousness, awareness of ourselves as subjective entities that are composed of one's physical being and those mental representations that accompany it.

Paradoxically, perhaps, although Buddhist thought maintains that the *skandhas* constitute an exhaustive catalog of what constitutes one's "self," it also claims that none of these aggregates—alone or in conjunction with others—constitutes a solid, identifiable, consistent unity of personhood. Not only are we unable to fully control any of these aggregates, but they are all necessarily impermanent (see the next chapter), which in turn distinguishes such a self from the imagined one—aka, the soul—that is ostensibly eternal and absolute,.

Buddhist thought explains our sense of personal continuity as due to *pratitya-samutpada*, often translated as "dependent arising," by which mental and physical events lead from one to another, a complex twelve-link chain of causality that goes from ignorance, through mental formations and consciousness, and ultimately to old age, death, and then (re)birth. (More of this, including a current, revised and somewhat biologized view, in chapter 4.)

Although intentionality is considered to be extremely important—it holds the key to whether an action has impact on one's karma (see chapter 6)—there is no underlying, primary ur-substance that remains steady and unchanged while all its secondary qualities (shape, color, size, strength, and so forth) vary over time. Emptiness—*anatman*—can also be described, especially in the language of Mahayana, as *sunya*, or "hollow," meaning that various entities—including but not exclusively human beings—are hollow inside, empty of their "own being." As Thich Nhat Hanh puts it, everything and everyone is composed of "not-self elements."

It is simply easier to say "I," "me," "you," and so on, than to refer to a particular combination of the five *skandhas*, just as a biologist will speak of a shark or a tuna rather than detailing the cells, tissues, and organs that interact with mutual dependence to generate each of the fishes in the sea. One more time: the fact that the Five Aggregates are empty when it comes to intrinsic existence does not mean that they literally aren't there. And similarly, let's note once again that if the being that possesses the Five Aggregates is human, then it's a person. Buddhism doesn't deny the commonsense idea of personhood, seen as a subjective psychological fact or even as an objective fact: something occupying three-dimensional space and typically perceived by another. Otherwise, the rather strict Buddhist prohibition against killing wouldn't make sense; as it is, killing is condemned by Buddhists because it violates the personhood of the victim.

In a dialogue reminiscent (to Western eyes) of the Socratic dialogues as recorded by Plato, the Buddha interrogates his followers:

"What do you think, monks? Are body…feeling…recognition…volitions…conscious awareness permanent or impermanent?"

"Impermanent, lord."

"But is something that is impermanent painful or unpainful?"

"Painful, lord."

"But is it fitting to regard something that is painful, whose nature it is to change as 'this is mine, I am this, this is my self'?"

"Certainly not, lord."

"Therefore, monks, all body…feeling…recognition…volitions…conscious awareness whatsoever, whether past, present or future, whether gross or subtle, inferior or refined, far or near, should be seen by means of clear understanding as it really is, as 'this is *not* mine, I am *not* this, this is *not* my self.'"[22]

Let us stipulate yet again that Buddhists are not fools. They realize that each of us experiences a degree of continuous memory as well as some continuity of body. They are not claiming that there is literally "no self," but rather that the various aggregates of personhood are not one's self. Buddhists certainly know the difference between the "you" who is reading these words and the "I" who has written them (or the other way around). But when we introspect—and also when we consider the objective biological facts—neither you nor I can find a solid, unchanging "me" aside from the feelings, sense data, ideas, and so forth that are apparent to our selves at any given moment.

The great philosopher David Hume—who, after all, was renowned as a hard-headed empiricist, not as a mystic or in any other way a devotee of Eastern thought—described in *A Treatise on Human Nature* the following widely shared experience, one that could as well have come from a Buddhist master:

> When I enter most intimately into what I call *myself*, I always stumble on some particular perception or other, of heat or cold, light or shade, love or hatred, pain or pleasure. I can never catch *myself* at any time without a perception, and never can observe any thing but the perception.

Hume goes on to note that when this ongoing progression of perceptions is ended, so is he.

There is an ancient Buddhist text, translated as "Milinda's Questions," that describes a conversation between a local Bactrian king, Milinda, and a monk, who, upon introducing himself as Nagasena, immediately corrects himself, saying that Nagasena is simply a label of convenience, which doesn't really refer to a distinct and identifiable person. Milinda responds as most of us would, accusing Nagasena of talking nonsense: " 'This Nagasena says there is no permanent individuality (no soul) implied in his name. Is it now even possible to approve him in that?' Turning to Nagasena, he said: 'If, most reverend Nagasena, there be no permanent individuality...who is it who lives a life of righteousness? Who is it who devotes himself to meditation?" Getting more personal (and perhaps even threatening), Milinda adds that, according to Nagasena's claim, "were a man to kill you, most reverend Nagasena, there would be no murder!"[23]

Milinda then goes through a remarkable list of body parts, asking whether Nagasena consists of any of them (I quote his inquiry as an example of the persistent tendency in Buddhist texts to be explicit and exhaustive...often exhausting as well):[24]

> Is it the nails, the teeth, the skin, the flesh, the nerves, the bones, the marrow, the kidneys, the heart, the liver, the abdomen, the spleen, the lungs, the larger intestines, the lower intestines, the stomach, the feces, the bile, the phlegm, the pus, the blood, the sweat, the fat, the tears, the serum, the saliva, the mucus, the oil that lubricates the joints, the urine, or the brain, or any or all of these, that is Nagasena?[v]

---

[v] Incidentally, this remarkable degree of detail is not unique to any one *sutra*; rather, it is characteristic of Buddhist stories and dialogs generally, testimony to its material, in-the-world specificity.

In response, the entity who goes by the name of Nagasena inquires how the king got to his hermitage, whereupon Milinda explains that he rode in a chariot. Nagasena then asks "Is it the axle that is the chariot?" "Certainly not." "Is it the wheels, or the framework, or the ropes, or the yoke, or the spokes of the wheels, or the goad, that are the chariot?" King Milinda denies that his chariot consists of any of its component parts. When asked if his chariot consists of anything *outside* its parts, he also agrees that it does not.

And yet, says Nagasena, the word "chariot" remains useful as a term of convenience, referring to something that depends upon a particular convergence of wheels, framework, ropes, and so on. By the same token, the name Nagasena is useful, but only as a verbal convenience: there is no fixed, independent, interior being by that or any other name. Insofar as there is coherence to Nagasena (or any other personal self), it is due simply to a connectedness of experience, a temporary aggregation of "not-self" elements, combined with verbal convenience.

There is a canonical Buddhist verse that goes as follows: "A beautiful woman, to a lover, is an object of desire. To a hermit, she is a distraction. To a wolf, a good meal." What, then, *is* the woman? Or any woman? Or man? Each of these things, and none. Perhaps this is too easy an evasion of common sense, because one could always point out that although the woman is different things to different observers, she nonetheless remains internally and foundationally herself. This is where the idea of "not-self" elements is crucial. Every person is different things to various observers, but in addition, she has no separate, inherent existence, independent of what makes her up, which is to say: aside from the rest of the world! We do not confront the world; rather, the world confronts itself, temporarily masquerading as each of us.

Buddhists have company, by the way, when it comes to this perplexing question of selfhood. Here is Alice, upon meeting a rather annoying Caterpillar in *Wonderland*:

> The Caterpillar and Alice looked at each other for some time in silence: at last the Caterpillar took the hookah out of its mouth, and addressed her in a languid, sleepy voice.
>
> "Who are *YOU*?" said the Caterpillar.
>
> This was not an encouraging opening for a conversation. Alice replied, rather shyly, "I–I hardly know, sir, just at present"—

As Buddhists as well as most biologists now recognize, the Caterpillar was actually asking a very deep question, such that Alice's befuddlement makes perfect sense.

Concerning the "self," the pioneering neurobiologist Charles Sherrington lamented that although our perceived individuality is actually a function of a near-infinity of complex, interacting parts (recall his metaphor of the "enchanted loom"), we are forced into a kind of Procrustean bed of unity:

> This self... regards itself as one, others treat it as one. It is addressed as one, by a name to which it answers. The Law and the State schedule it as one. They identify it with a body which is considered by it and them to belong to it integrally. In short, unchallenged and untargeted conviction assumes it to be one. The logic of grammar endorses this by a pronoun in the singular. All its diversity is merged in oneness.[25]

What we perceive as our self is necessitated by our biological need to draw conceptual boundaries between our own bodies, our genome, and the cultural and social system in which we exist. But at the same time, insofar as we are necessarily dependent on other people, other beings, other systems, we cannot be altogether separate and distinct. "Each healthy living cell in your body controls its own activity," writes Thich Nhat Hanh, "but does this mean that each cell has its own self? In the biological classification system, species make up smaller subdivisions of genus. Does each species represent a 'self'?"[26]

Modern evolutionary biology lends yet more instability to the seemingly coherent self. In the evocative (and biologically accurate) terminology of Richard Dawkins, organisms are "survival machines" constructed by their genes for the express purpose of getting those genes projected into the future.[27] As such, these machines are temporary: we have already noted that life spans are finite and generally quite short, with only genes having any prospect of long-term persistence (although, as we shall see in the next chapter, they too are doomed to change, via mutation). Moreover, the exact boundaries of these survival machines—aka bodies, or organisms, or you, or me—are themselves indistinct, and not simply because we have a hard time discerning those boundaries but because they really are indistinct.

Consider the beaver, buck teeth, lush fur, flat tail, and all. It seems easy enough to designate the outlines of its beaverness. Its heart is included; ditto for its lungs, stomach, and intestines, as well as its more obvious external structures. But does that furry beaver pelt constitute the genuine outer boundary of the animal, and—more to the point—of its underlying DNA? After all, the beaver's body is identifiable to us because it is a discernible manifestation of its underlying genome, which, acting in conjunction with each animal's environmental experience, produces what biologists call its particular phenotype—those aspects of a living thing that are immediately detectable by our sense organs, how it looks, sounds, feels, smells, and so forth.

But what about the beaver's dam, or its lodge? In a very real sense, these too have been produced by the animal's genotype, acting in concert with its surroundings, which has led Dawkins to call beaver dams and lodges, birds' nests, spiderwebs, and so forth examples of each creature's "extended phenotype." Such things exemplify the "long reach of the gene,"[28] and in the process, the indeterminate nature of every organism's self.

Our own sense of self—like our sense of what legitimately constitutes a beaver—is similarly built on shifting sand and thus bound to be unstable. Our narrative of personhood as independent and separate is inherently fragile. We keep trying to build up a self-identity, but since this exists only in the context of other things as well as other organisms—whose boundaries are equally indistinct—it never lasts for long, nor is it very robust.[vi]

Nonetheless, absent a Buddhistic appreciation of "not-self," most people have little doubt where the buck stops. John Donne be damned: each of us, we typically assume, is an island, entire unto him or herself, everyone his own man or woman, separate, distinct, independent, and in charge. An army of one. At least, this is what cultural tradition and subjective experience tell us.

But let's add a bit more biology—in this case, parasitology—and go inward in our search for the paradox of selflessness, instead of outward, as we just did when examining the notion of extended phenotypes. Consider the disconcerting fact that there are many more freeloaders in the world than free-livers, many more parasites and pathogens than individual, stand-alone organisms. After all, pretty much every multicellular animal is home to numerous fellow travelers, and—this is the point— each of these creatures has, in a sense, its own agenda. Considering just one group of worms, invertebrate biologist Ralph Buchsbaum suggested that "if all the matter in the universe except the nematodes were swept away, our world would still be dimly recognizable.... Trees would still stand in ghostly rows representing our streets and highways. The location of the various plants and animals would still be decipherable, and, had we sufficient knowledge, in many cases even their species could be determined by an examination of their erstwhile nematode parasites."[29]

Put another way, if a biologist were to answer Frank Zappa's acid-rock question, "Suzy Creamcheese...what's got into you?" the response might well be: "A whole lot of other living things."

What difference does this make? For many of us supposedly free-living creatures, quite a lot. Providing room and board to other life forms not only compromises one's nutritional status (not to mention peace of mind), it often reduces freedom

---

[vi] It is an interesting and—so far as I know—unanswered question whether the human sense of self becomes more or less stable over time during the personal trajectory we describe as aging.

of action, too. The technical phrase is "host manipulation." Case in point: the tapeworm *Echinococcus multilocularus* causes its mouse host to become obese and sluggish, making it easy pickings for predators, notably foxes which—not coincidentally—constitute the next phase in the tapeworm's life cycle. Those the gods intend to bring low, according to the Greeks, they first make proud. Those tapeworms intending to migrate from mouse to fox do so by first making "their" mouse fat, sluggish, and, thus, fox food.

Sometimes the process is more bizarre. For example, the life cycle of a trematode worm known as *Dicrocoelium dentriticum* involves doing time inside a snail, after which the worms are shed inside slime balls, which are then eaten by an ant; the ant is then eaten by a sheep. Getting from its insect host (ant) to its mammalian one (sheep) is a bit of a stretch, but the resourceful worm has found a way: ensconced within their "host" ant, some of the worms become, well, antsy. They proceed to migrate to its formicine brain, whereupon they manage to rewire their host's neurons and hijack its behavior. The manipulated ant, acting with zombie-like fidelity to *Dicrocoelium's* demands, climbs to the top of a blade of grass and clamps down with its jaws, whereupon it waits patiently and conspicuously until it is consumed by a grazing sheep. Thus transported to its desired happy breeding ground deep inside sheep bowels, the worm turns and releases its eggs, which depart with a healthy helping of sheep poop, only to be consumed once more by snails and then by ants. It's a distressingly frequent story—distressing, at least, to those committed to autonomy.

A final example, as unappetizing as it is important: plague, the notorious Black Death, is caused by bacteria carried by fleas, which, in turn, live mostly on rats. Rat fleas sup cheerfully on rat blood but will happily nibble people, too, and when they are infected with the plague bacillus, they spread the illness from rat to human. The important point for our purposes is that once they are infected with plague, disease-ridden fleas are especially enthusiastic diners, because the plague bacillus multiplies within flea stomachs, diabolically rendering the tiny insects incapable of satisfying their growing hunger. Not only are these fleas especially voracious in their frustration, but because bacteria are cramming its own belly, an infected flea vomits blood back into the wound it has just made, introducing plague bacilli into yet another victim. A desperately hungry, frustrated, plague-promoting flea, if asked, might well claim that "the devil made me do it," but in fact, it is the handiwork of *Pasteurellis pestis*. ("So, naturalists observe, a flea has smaller fleas that on him prey," wrote Jonathan Swift in *On Poetry: a rhapsody*, in 1733. "And these have smaller still to bite 'em. And so proceed ad infinitum.")

Not that a plague bacterium—any more than a mouse-dwelling tapeworm or ant-hijacking brain worm—knows what it is doing when it reorders the inclinations of its host. Rather, a long evolutionary history has arranged things so that the manipulators have inherited the earth.

Of course, things are not quite this simple. The ways of natural selection are devious and deep, embracing not only would-be manipulators but also their intended victims. Hosts needn't meekly follow just because others seek to lead. In Shakespeare's *Henry IV, Part I*, Owen Glendower boasts, "I can call spirits from the vasty deep," to which Hotspur responds, "Why, so can I or so can any man; But will they come when you do call for them?" Sometimes it is unclear whether our spirits, no less than those Glendower might seek to enlist, act at the behest of others or themselves. Take coughing, or sneezing, or even—since we have already broached some indelicate matters—diarrhea.

When people get sick, they often cough and sneeze. Indeed, aside from feeling crummy or possibly running a fever, coughing and sneezing are important ways we identify being ill in the first place. It may be beneficial for an infected person to cough up and sneeze out some of the tiny organismic invaders, although, to be sure, it isn't so beneficial for others nearby. This, in turn, leads to an interesting possibility: what if coughing and sneezing aren't merely symptoms of disease, but also— even primarily—a manipulation of us, the host, by, say, the influenza virus? Shades of fattened mice and grass blade–besotted ants. As to diarrhea, consider that it is a major (and potentially deadly) consequence of cholera. To be sure, as with a flu victim's sneezing and coughing, perhaps it benefits a cholera sufferer to expel the cholera-causing *Vibrio cholerae*. But it also benefits *Vibrio cholerae*. Just as Lenin urged us to ask, "Who, whom?" with regard to social interactions—who benefits at the expense of whom?—an evolutionary perspective urges upon us the wisdom of asking a similar question. Who benefits when a cholera victim "shits his guts out" and dies? The cholera bacillus.

This most dramatic symptom of cholera is caused by a toxin produced by the bacillus, making the host's intestines permeable to water, which gushes into the gut in vast quantities, making for a colonic flood that washes out much of the native bacterial intestinal flora, leaving a comparatively competitor-free environment in which *Vibrio cholerae* can flourish. A big part of that flourishing, moreover, occurs via flushing, with more than 100 billion *V. cholerae* per liter of effluent sluicing out of the victim's body, whereupon, if conditions are less than hygienic, they can infect new victims. Diarrhea, then, isn't just a symptom of cholera: it is a successful manipulation of *Homo sapiens* by and for the bacteria.

Intriguing as these tales of pathology may be, it is too easy to shrug them off when it comes to the daily lives most of us experience. After all, aside from sneezing, coughing, or pooping, our actions are, we like to insist, ours and ours alone, if only because we see ourselves as acting on our own volition and not for the benefit of some parasitic or pathogenic occupying army. So when we fall in love, we do so for ourselves, not at the behest of a romance-addled tapeworm. When we help a friend, we aren't being manipulated by an altruistic bacterium. If we eat when hungry, sleep when tired, scratch an itch, or write a poem, we aren't knuckling under to the needs of our nematodes. But in fact, there's more going on here than is immediately obvious—and for some people, more than they would like.

Think about having a child. Not how it feels to become a parent, what it costs, or the social, family, personal, or cultural factors involved. Rather, think about it as Lenin suggested (who, whom?), or as a modern-day Darwinian might: who—or rather, what—benefits from the act of reproduction? Return to the ruling mantra of revolutionary biology: it's the genes, stupid. They're the beneficiaries of baby making, the reason for reproducing, just as they are the reason for kin-based altruism. As modern evolutionary biologists increasingly recognize, bodies—more to the point, babies—are our genes' way of projecting themselves into the future. Or as Dawkins has so effectively argued, bodies are temporary, ephemeral, short-lived survival vehicles for those genes, which are the only entities that persist over evolutionary time. No matter how much money, time, or effort is lavished on them, regardless of how much they are exercised, pampered, monitored for bad cholesterol, or modified via botox injections or plastic surgery, bodies—as we have seen—don't have much of a future.

And so, to return to the key questions, "Who's in charge?" "Who's calling and who's heeding?" the biologically and Buddhistically informed answer is not all that different from those alarming rat/tapeworm, ant/trematode, flea/bacteria relationships, only this time it's genes/body. Unlike parasites or pathogens, when it comes to genes manipulating their "host" bodies, the situation seems less dire to contemplate, if only because it is less a matter of demonic possession than of our genes, our selves. The problem, however, is that those presumably personal genes aren't any more hesitant about manipulating our selves than is a brain worm about hijacking an ant.

Take a seemingly more benign behavior, indeed, one that is highly esteemed: altruism. This is a favorite of evolutionary biologists, because superficially, every altruistic act is a paradox. Natural selection should squash any genetically mediated tendency to confer benefits on someone else while disadvantaging the altruist. Such genes ought, therefore, to disappear from the gene pool, to be replaced by their more selfish counterparts. To a large extent, however, the paradox of altruism has been resolved by the recognition that "selfish genes"—in the sense of Dawkins and others—can promote themselves (rather, identical copies of themselves), by conferring benefits on genetic relatives, who are likely to carry copies of the genes in question.

By this process, known as "kin selection," behavior that appears altruistic at the level of bodies is revealed to be selfish at the level of genes. Nepotism is natural. (So, by the way, is gangrene; natural and "good" aren't necessarily the same.) When someone favors a genetic relative, who then is doing the favoring, the nepotist or the nepotist's genes? Just as sneezing may well be a successful manipulation of us (*Homo sapiens*) by them (viruses), what about altruism as another successful manipulation of us, this time by our own altruism genes? Admirable as altruism may be, it is therefore, in a sense, yet another form of manipulation, with the manipulated victim—the altruist—acting at the behest of some of his or her own genes. After all, just as the brain worm gains by orchestrating the actions of an ant, altruism genes stand to gain when we are nice to our cousin Sarah. Never mind that such niceness is by definition costly for the helper. It's really a gift from altruism genes within the giver to altruism genes within the recipient.

All this may seem a bit naive, since biologists know that genes don't order their bodies around. No characteristic of any living thing emerges full-grown from the coils of DNA, like Athena springing out of the forehead of Zeus. As we have already emphasized, every phenotypic trait—including behavior—results from a complex interaction of genetic potential and experience, learning as well as instinct, nurture inextricably combined with nature. Life is a matter of genetic influence, not determinism.

But does this really resolve the problem? Let's say that a brain worm–bearing ant still possesses some free will, and that a trematode-carrying mouse has even a bit more. What if their behavior was influenced but not determined? Wouldn't even influence be enough to cast doubt on their agency, their independence of action? And (here comes at least one Big Question), why should human agency or free will be any less suspect? Even if we are manipulated just a tiny bit by our genes, isn't that enough to raise once again that disconcerting question: who's in charge here?

Maybe it doesn't matter. Or, put differently, and once again in ways that a Buddhist master would confirm, maybe there is no one in charge. That is, no one distinguishable from everyone and everything else. If so, then this is because the environment is no more outside us than inside, part tapeworm, part bacterium, part genes, and—once again—no independent, self-serving, order-issuing homunculus. We are manipulated by, no less than manipulators of, the rest of life inside and all around us. And so, the answer to "who's in charge here?" depends on what the meaning of "who" is. Who is left after the parasites and pathogens are removed? And after you are separated from your genes?

Biologists and physicians often confront other, comparable issues of self versus not-self, notably when it comes to immune responses and the body's natural tendency to reject foreign tissue. This, in turn, requires that in the case of organ transplants, various drugs be used to eliminate—or at least to blunt—such rejection responses. Interesting problems also arise when it comes to defining the literal limits of individuality. What, for example, should we make of clones—artificial as promised (or threatened) by advances in biotechnology, as well as natural such as in the case of aspen groves, which, although they appear to consist of individual trees, are actually one "individual," connected via underground root systems?

Then there are those unicellular organisms that reproduce by simply dividing, as occurs among amoebas: we generally consider that once the cell membrane that had comprised one amoeba has pinched off, thereby leaving two daughter cells, the amoeba has lost its individuality. Yet, when cells that eventually make up a human body reproduce in precisely the same way—via mitosis—they remain somehow part of the larger whole that we insist (somewhat perversely and un-Buddhistically) on labeling as a single "individual."

The situation is even more fraught. Consider, for example, the curious fact that within each healthy human being there are about 100 trillion cells, most of them occupying the small and large intestine, that are literally foreign in that they are microbes with a completely different genome from that of the body they inhabit. Yet they aren't parasites or pathogens; it turns out, in fact, that they are crucial to health, involved not only in digesting nutrients but in generating appropriate immune responses, producing a variety of necessary enzymes, neurotransmitters, and various biochemical substances while doubtless serving other functions not yet identified. Taken as a whole, each person harbors about ten times as many foreign cells as it does those personal ones with the genetic signature associated in most of our minds with one's "self."

Given all this perplexity, let's leave the last words to a modern icon of organic, oceanic wisdom, SpongeBob SquarePants, a fictional character recognizable by many children and more than a few adults. Mr. SquarePants is a cheerful, talkative—although admittedly, somewhat cartoonish—fellow of the phylum Porifera, who, according to his theme song, "lives in a pineapple under the sea…Absorbent and yellow and porous is he." I don't know about the pineapple or the yellow, but absorbent and porous are we, too.

3

Impermanence (*Anitya*)

"Things fall apart," wrote W. B. Yeats in his poem *The Second Coming* (1919); "the center cannot hold / Mere anarchy is loosed upon the world."

Biologists understand that entropy inevitably increases, which is why the maintenance of living systems requires a constant input of energy: photons from the sun for plants, which ultimately provides food for all animals, too. The important ecological concept of "trophic levels" derives from this inevitable tendency of matter to "run down." Thus, it takes, for example, about a thousand pounds of grass to make one hundred pounds of cow to make ten pounds of person to make one pound of vulture. People, too, "run down," become ill, and senesce, as do cows and vultures, even grass.

The Buddha is said to have been deeply affected when he witnessed different people suffering, in turn, from illness, old age, and finally, death. The ensuing realization—the unavoidability of structural breakdown—is another way of posing the problem of what Newton developed as his Second Law of Thermodynamics: things fall apart, yielding inevitable pain and suffering. It is a domain about which Buddhist insight is especially cogent. Indeed, the Four Noble Truths of Buddhism are concerned entirely with suffering: the first three identify its existence and explore its causation; the fourth, the Eightfold Path, describes how suffering can be alleviated. In Mahayana Buddhist tradition, *boddhisattvas* are a kind of social and ecological conscience keepers, individuals dedicated to enlightenment and to relieving the suffering of all beings. For their part, ecologists increasingly perceive the shared pain implied in life's shared reality, and in the breakdown of ecosystems and natural communities. There is also a lively discussion among evolutionary biologists and philosophers about whether ethical principles can or should be derived from the workings of natural selection, a process that operates without regard to the suffering it often mandates.

For Buddhists, the origin of suffering is to be found in excessive and inappropriate desires (*tanha*), especially a tendency to cling to things that are unachievable, undesirable, and/or—pertinent to the current chapter—necessarily perishable. And so we come to the idea—and the reality—of impermanence.

In the world before Dolly (the first artificially cloned sheep) it was believed that cellular differentiation was a one-way street, and that once development proceeds beyond the earliest cell divisions, the fate of every cell is rigidly predetermined. Now we know better. Recall that Dolly was produced when a mammary gland cell was essentially vivisected in a laboratory and its nucleus implanted into an egg cell whose own nucleus had been removed.[i] As a result, a new sheep was produced whose genotype, contained in the DNA within the transplanted nucleus, was a clone, identical in genetic makeup to its "mother."

A take-home message, therefore, from Dolly the sheep is that cells don't necessarily remain permanently what they had been, since the nucleus of a fully differentiated mammary cell revealed itself capable of giving rise to an entire sheep. The capacity for change, flexibility, and impermanence is—ironically—built into life itself. (One is tempted, in fact, to point out the paradox that this capacity is itself permanent!) Stem-cell research shows, moreover, that much human tissue is also multipotent, capable of differentiating in a variety of ways. This is but one example of how the latest developments in biology converge upon some of the oldest realizations of Buddhism: in this case, the underlying truth of impermanence—in Sanskrit, *anitya*.

Along these same lines, whole genomes are currently permeable to the introduction of selected genes from other lineages (freeze-resistant halibut genes into strawberry plants, and so on), raising possibilities—and fears—about a future in which the boundaries of cell, individual, and species become increasingly vague. It isn't simply that living things have the capacity for impermanence, but rather that they in fact are shape-shifters at their core.

One of the more startling discoveries in cell biology has also been the widespread acceptance of what had long been considered a fringe idea: so-called endosymbiotic theory. This idea, which is especially associated with the work of Lynn Margulis (a biologist who was also the first wife of the astronomer Carl Sagan), provides a novel

---

[i] It was because this animal was derived from a mammary gland cell that she was named "Dolly," after Dolly Parton, also known—even among wonky lab scientists—for her mammaries.

explanation for the evolution of eukaryotic cells, those fundamental components of life that contain nuclei along with other crucial internal parts such as mitochondria (responsible for most of a cell's energy production) and chloroplasts (which do the work of photosynthesis in plants).

According to endosymbiotic theory, these various internal cell organelles originated as free-living, primitive bacteria that were "eaten" by other cells, after which they eventually established a cooperative, symbiotic relationship. Mitochondria were originally aerobic protobacteria, of the sort that are currently common in modern oceans. Once inside another cell—which was initially anaerobic (unable to use oxygen to break down potential food molecules)—these newly constructed oxygen-using bacteria were able to thrive, using the partly digested organic molecules already present within their host, material that had not been metabolically available to the ancestral, anaerobic host cell. Both the original, anaerobic, predatory cell and its smaller aerobic, bacterial victim were better off, thus giving rise over time[ii] to modern cells, complete with their energy-providing mitochondria. And these cells, in turn, eventually came together to produce multicellular creatures, including human beings.

A similar story seems likely with respect to the evolution of chloroplasts, if a primitive bacterium capable of photosynthesis was engulfed but not digestively destroyed by a larger, nonphotosynthetic cell. Once again, over time, both "predator" and "prey" could have settled down to a cooperative, mutually beneficial evolutionary trajectory, eventually producing what we call plants. The various tiny subcellular victims (which might well have included primitive flagella, microtubules, and other components of modern cells) would have eventually become happy colonizers and then, with time, essential components of their hosts. This theory gained strong support when it was found that the DNA of such endosymbionts was quite different from that of the host's nuclear DNA.

This theory, if correct, has important implications for the role of cooperation versus competition in the evolution of life itself. But for our purposes, it is also significant as yet another example of how even the fundamental building blocks of life—namely, cells—are best understood as the result of shifting, changing, interdependent dynamics.

Although new to biology, these and other developments were long ago foreshadowed in Buddhist thought, notably by *anitya*. Not surprisingly, this concept

---

ii One thing that evolutionary processes generally require—and have in abundance—is time. And there is nothing like time to generate, eventually, impermanence!

is partly rooted in the recognition of not-self (*anatman*), described in chapter 2; in conjunction with *anatman*, *anitya* gives rise to connectedness (*pratitya-samutpada*, which we shall explore in the next chapter). Just as Tibetan Buddhist sand mandalas are typically designed to be destroyed—as a reminder of our own impermanence— Buddhist doctrine recognizes that nothing is fixed or irreversible; everything is in a state of flux, brimming with uncertainty, inconstancy, and (to put a more positive spin on all this) possibilities.

One of the key concepts in physiology is homeostasis, the capacity of living things to maintain their internal selves within a remarkably narrow range of variability. Thus, mammals especially are famous for maintaining their internal body temperatures within narrow limits: too hot and they have a fever, too cold and they are vulnerable to hypothermia. By a complex system of biochemical thermostats, all animals similarly keep their internal fluids—notably but not limited to their blood—in a continuing Goldilocks state: not too acidic, not too alkaline, not too salty, not too dilute, and so on. To deviate from these and other crucial parameters, even by a very small amount, is certain death.

But even here, in what appears to be a profound case of maintaining permanence and stability rather than impermanence and change, biological systems drift strongly toward *anitya*. Although one of the defining characteristics of living things is their constancy—their defiance of the Second Law of Thermodynamics, which mandates that order decreases and disorder (entropy) inevitably increases over time in the case of closed systems—in fact a *sine qua non* of life is that every living system must be an open one: to be alive is to engage in a constant ebb and flow whereby the organism obtains necessary energy from its surroundings. In the case of green plants, this involves capturing photons and converting them into energy-rich, organic molecules, while for animals it requires living off the bounty of those plants, or second (sometimes even third) hand, by consuming animals who themselves have consumed plants. Paradoxically, therefore, even the maintenance of constancy requires constant openness to change,[iii] in this case *ex*change with the organism's environment.

To reiterate: there is simply no life that is based on steady-state, fixed, and frozen relations with its surroundings. Rather, homeostasis—prerequisite for being alive— is achieved only by those particular aggregations of matter and energy that we designate organisms or living things or bodies, whose primary defining characteristic is that they are open to the rest of the world, from which they derive energy in order to continue. Permanence is a death sentence.

[iii] And of course, no living things remain constant in their aliveness; the inevitability of death is prominent in Buddhist thought.

In basing a thought tradition on impermanence, Buddhism is unique among the world's great religions, all of which tend to focus instead on that which is supposed to be unchanging and eternal, especially souls and, even more, God. By contrast, for Buddhists, everything is changing, temporary, dependent on other things. Living creatures in particular—for Buddhists and biologists—are the literal embodiment of *anitya*.

Question: when a Buddhist nun goes to a beauty parlor, what is she is likely to get? Answer: an impermanent.

More seriously, the Buddhist approach is in many ways an antidote to Plato, who equated the real with the timeless and unchanging, and in that sense generated a philosophy that was especially convivial to the later rise of Christianity. According to Plato's philosophy, the material world is necessarily flawed and imperfect, in comparison with the "real" existence of idealized forms. A chair, for example, is necessarily imperfect, being a bit off-kilter, having dents, dings, and the like; by contrast, the ideal form of a chair persists in the imagination of the carpenter and is unchanging. Such a viewpoint leads, in turn, to the conclusion that the world is less than real, something to be disregarded or regretted, while one strives to attain perfect, absolute, unchanging Truth.

To some extent, this is what the young Buddha tried to do after he encountered the realities of sickness, old age, and death: to go beyond and through the painful realities of life, via asceticism and mortification of the flesh. Eventually, we are told, after starving himself almost to the point of suicide, he came to enlightenment, part of which involved the recognition that all things are ever-changing and impermanent. Unlike Christ, who promises eternal life, the last words of the Buddha are reported to have begun with the observation, "Decay is inherent in all things."

As my favorite Buddhist, Thich Nhat Hanh, puts it, impermanence is intimately tied to continuity:

We cannot conceive of the birth of anything. There is only continuation.... Look back further and you will see that you not only exist in your father and mother, but you also exist in your grandparents and in your great grandparents.... I know in the past I have been a cloud, a river, and the air.... This is the history of life on earth. We have been gas, sunshine, water, fungi, and plants.... Nothing can be born and also nothing can die.[1]

There has been surprisingly little formal exchange between Buddhism and biology. Most biologists—at least, those in the West—tend to operate under the typically unspoken assumption that things lack *anatman* and *anitya*—that is, that the objects of their study are to a large extent self-contained and permanent (or at least, long-lasting), even though increasingly their findings point in the other direction. This disconnect may be due to an unacknowledged impact of the Judeo-Christian religious tradition, which, as we have noted, emphasizes the persistence of certain eternal verities. Some blame can also be laid at the foot of simple "common sense," which we all know can be misleading and almost certainly is in this case.

At any given moment in time, we think we see things as they are, although in fact we are merely capturing a flickering moment in time. A bit more perspective, the equivalent of fast-motion photography, shows living things coming into existence, growing, achieving adulthood, then decreasing and deteriorating, their components recycled into new organisms, and so forth. As a result, we tend to see these differ-ent states as mutually contradictory. A living creature cannot be simultaneously an embryo, a juvenile, an adult, a senescent individual, and a mass of disordered tissues and molecules. But in fact, any such thing is in a state of flux, of continual coming and going; it is never constant, immutable, and solid. This is so fundamental to mod-ern physics that it hardly needs restating. A block of wood, we know for certain, is mostly empty space, and those parts made up of solid subatomic particles either are moving with unimaginable rapidity or "exist" as clouds of probability rather than as stationary, reliable objects.

Just as it is misleading for Westerners to think that "not-self" means that you don't really exist (when each of us knows darn well that we do!), don't mistake *anitya* for a kind of amorphous, unpredictable variability. Rather, just as *anatman* refers to things' being empty of any claim to solitary, independent, atomistic, disconnected, and isolated existence, *anitya* does not argue that things are literally so changeable as to be lacking in identifiable form. *Anitya* is simply another way of saying that nothing stays the same from one moment to the next. As Yeats both observed and inquired in *Among School Children* (1926):

Ah, body swayed to music, ah, brightening glance
how can we know the dancer from the dance?

The world of living things is also impermanent—only more so. Thus, not only are organisms composed of the same ever-changing matter as are inorganic entities, they exhibit an extraordinary degree of change even when seen on a larger scale, as whole beings. Dancers themselves, they are inseparable from their dance. There was a time, however, when the living and nonliving were thought to occupy different planes of existence.

It is difficult for us, in the twenty-first century, to appreciate the scientific and philosophical impact of the discovery by Friedrich Wöhler, in 1828, that he could convert ammonium cyanate, an inorganic molecule, into urea. Prior to this, it was believed that organic chemicals such as urea were uniquely associated with life itself and fundamentally distinct from nonliving, inorganic components. Urea had been discovered in 1799 and prior to its synthesis could be obtained only from the urine of living organisms. The ability to construct this substance from nonliving components not only gave rise to the new discipline of organic chemistry, it also disproved the prevailing theory of "vitalism," that living things were characterized by a unique *élan vital* (vital force) that existed independent of its material components.[iv]

Now we know for certain that living stuff is constantly replenished by, and created from, nonliving components. The point here is close to the old advertising jingle for milk: "There's a new you coming every day." Except that it's more like every hour, minute, instant. Our present existence takes place only on the instantaneous knife-edge of now: while you were reading these words, you changed—not only irrevocably, but many times.

Let's return to "our bodies, ourselves," something that isn't true for long and not in the way that most people think. As we've already seen, there is also room to question exactly who this "person" is who constitutes that body. In any event, it isn't the body that's illusory, but our insistence that it is permanent, persistent, separate, and consistent over time. The embodied self—just like the ostensibly conscious, little-green-man homuncular self inside our heads—is a convenient fiction. Buddhist teaching calls it *svabhava*, the supposedly fixed or unchanging essence, about which so many have been fooled for so long. We previously noted that for Norbert Wiener, famous for elucidating the mathematics of systems theory, "We are not stuff that abides, we are patterns...in a river of ever-flowing water."[2]

Another metaphor, often used by Buddhist thinkers, is similarly aquatic: we are like the eddies in a river, readily identified and yet, more than impermanent; rather, we are constantly moving, ever constituted and then reconstituted of materials that pass through a particular pattern that we identify as "our self," yet never really the same from one moment to the next.

Yet, oddly, there is a sense in which compared with the traditional Western standpoint, a Buddhist view reveals greater permanence, not less (at least in comparison to my own atheist perspective, in which death is the end). Thus, for Buddhists, death isn't the end of life—just the end of *this* life. We can't really die because we were never really born, just recycled, and are therefore always in the act of becoming. As already noted, life itself requires metabolism, the constant taking in of material that

---

[iv] Describing his discovery, Wöhler wrote in understandable excitement to his mentor, the chemist Jöns Jacob Berzelius, that "I cannot, so to say, hold my chemical water and must tell you that I can make urea without thereby needing to have kidneys, or anyhow, an animal, be it human or dog."

must be broken down and its energy released and then captured via cell metabolism in order to maintain this state of flux that we identify as being alive.

Regular material and energetic "refreshment" is in fact key to life and is best known even to nonbiologists as respiration, as well as digestion and metabolism. During respiration, the average person takes in (inhales) about half a liter of air, relatively rich in oxygen and low in carbon dioxide, and then breathes out (exhales) that same half a liter, which now contains less oxygen and more carbon dioxide. Similarly, we eat not simply because we get hungry, but rather, we get hungry because we need to take in energy-rich material (aka food); this is then broken down in complex ways and eventually integrated into our bodily substance, whereupon it passes into each living cell to provide nourishment for that cell, aided by the oxygen obtained via respiration. At this point, it is tempting to regale the reader with specific numbers describing the volume of new oxygen atoms incorporated into the body every minute, as well as the quantity of carbon, hydrogen, phosphorus, sulfur, potassium, calcium, and so forth that is similarly processed in order to keep that body in a dynamic state that we refer to as "alive." Instead of doing so, however, let's simply note that this continuing input and output involves a nearly complete recycling of our constituent atoms every few days.

And not just ours. The Zen-influenced, late-seventeenth-century Japanese haiku master Basho wrote a brief masterpiece as follows: "A roadside hibiscus: / My horse ate it." Let's deconstruct this poem, along with the flower as well as the horse: following its momentous encounter, Basho's horse became somewhat hibiscus, and the hibiscus, a component of Basho's horse. In the process, the hibiscus ceased to "be" what it had been, as it "became" part horse. But meditate on this a bit more. Before its encounter with Basho's hungry horse, did that hibiscus exist as a separate, distinct, self-contained object? In its seemingly hibiscusoid state, wasn't it composed of carbon, hydrogen, phosphorus, sulfur, and so forth—that is to say, various nonhibiscus elements? And once incorporated into the horse, weren't those nonhibiscus substances simply rearranged and reconfigured (albeit in complex ways), so as ultimately to be recognized by the rest of us—Basho included—as something superficially different but nonetheless equally alive?

This condition of aliveness is not only precious, it is also temporary and perhaps all the more valuable *because* it is temporary. It is also vulnerable to disruption from what we customarily but inaccurately refer as the "outside": injury, attack by predators, pathogens, sharp sticks, heavy rocks, hungry horses, or clever coyotes, and also (the most serious threat of all) susceptible to running down if it isn't regularly supplied with a continuing stream of elemental atoms as well as energy-containing compounds. As Thich Nhat Hanh emphasized, we are all made of not-self elements that have temporarily been corralled into what we like to call our bodies. Nor are these atoms and molecules imbued with anything mystical. They're just atoms and molecules.

Despite the urea-creating breakthrough of Friedrich Wöhler and the entirety of organic chemistry as well as the burgeoning science of physiology, the myth that living things are qualitatively discontinuous from the nonliving has been stubbornly persistent. Thus, the influential French philosopher Henri Bergson wrote that organisms were alive by virtue of their *élan vital*, which, critics promptly noted, was like explaining the movement of a train by noting that it contained *élan locomotif*. The important thing about the continuing flow of materials that enable and literally embody life is that they—the atoms of oxygen, carbon, hydrogen, and so on, as well as the molecules of carbohydrate, amino acids, fats, salts, and so forth—are not alive in themselves. They are no different from those same atoms and molecules present in rocks or soil. What generates aliveness is the particular arrangement and organization of these substances, arrangements and organizations that are constantly pulsing, breathing, respiring, digesting, becoming, living, dying, decaying, reforming, and so forth.

In addition, recall the astounding fact—described in the preceding chapter—that a human being contains approximately ten times more "foreign" than "self" cells, an observation that, as we have seen, buttresses the claim of *anatman* (not-self). Consider, as well, that when a person is born, he or she contains virtually none of these foreign cells; they are obtained during the course of living, including, for example, a benevolent dosing of anaerobic microbes from the mother as the emerging infant traverses its mother's birth canal.[v] As a result, we are helpfully infected with a burgeoning microbial community that quickly comes to outnumber our own constituent cells. Thus, we go from being mostly ourselves as newborns to being mostly not-self elements, a transition that is nothing if not *anitya*—even when it comes to our innermost being.

Recall Alice's disconcerting encounter with the Caterpillar, in which Alice confessed to not knowing who she was. As their conversation continued, it became clear that Alice's confusion derived specifically from the numerous and troubling changes she had been experiencing:

"At least I know who I *WAS* when I got up this morning, but I think I must have been changed several times since then."

"What do you mean by that?" said the Caterpillar sternly. "Explain yourself!"

---

[v] In itself, this crucial maternal bequeathal to her newborn is enough to justify reducing the number of cesarean sections to a minimum.

"I can't explain *MYSELF*, I'm afraid, sir," said Alice, "because I'm not myself, you see."

"I don't see," said the Caterpillar.

"I'm afraid I can't put it more clearly," Alice replied very politely, "for I can't understand it myself to begin with; and being so many different sizes in a day is very confusing."

"It isn't," said the Caterpillar.

"Well, perhaps you haven't found it so yet," said Alice; "but when you have to turn into a chrysalis–you will some day, you know–and then after that into a butterfly, I should think you'll feel it a little queer, won't you?"

"Not a bit," said the Caterpillar.

"Well, perhaps your feelings may be different," said Alice; "all I know is, it would feel very queer to *ME*."

"You!" said the Caterpillar contemptuously. "Who are YOU?"

Which brought them back again to the beginning of the conversation. Alice felt a little irritated at the Caterpillar's making such *VERY* short remarks, and she drew herself up and said, very gravely, "I think, you ought to tell me who *YOU* are, first."

Excellent question, young lady!

When it comes to evolutionary biology in general and genes in particular, it would seem that in the realm of DNA we encounter something that is lacking in *anitya*—or, to sound more positive, life-related entities that are permanent rather than temporary. After all, one of the most important recent realizations on the part of biologists has been that although bodies are temporary and impermanent, their constituent genes are potentially immortal, what we might dub anti-*anitya*.

Indeed, Richard Dawkins employed the phrase "Immortal Coils" when titling one of the chapters of his now-classic exposition *The Selfish Gene*.[vi] His point—and one that is endorsed by essentially all evolutionary biologists—is that genes are typically passed unchanged from generation to generation. As Thomas Hardy put it in the poem reproduced in chapter 2, they "heed no call to die."

But they do change.

One of the wonders of living things is the extraordinary fidelity with which DNA molecules manage to copy themselves with every cell division. Geneticists are keenly

---

[vi] A play on Shakespeare's description of the body as a "mortal coil," to be shuffled off at death.

aware that some genes—typically those underlying structures or functions that are especially important to an organism's survival and reproduction—are "highly conserved," which simply means that they change very little from one generation to the next. And yet, although they may seem permanent when observed and studied at any given point in time, the reality is that all genes partake of *anitya*—slowly, to be sure, but ineluctably. A very rough estimate is that on average, genetic mutations arise about one time in a million replications. Even though such events are essentially copying errors, nearly always resulting in cells with lower biological fitness than their unmutated cousins, mutations are ultimately fundamental to the evolutionary process, since this is how DNA diversity is generated. In a nonmutating world, and especially if this world also lacked sexual recombination (the other major source of genetic variation), individuals would be genetically identical and evolutionary change would thus be stymied.

Don't get the impression, incidentally, that just because they are fundamental to the evolutionary process, mutations are necessarily part of some grand cosmic plan aiming toward evolution—especially ours—as an endpoint. Rather, mutations are simply an inevitable consequence of how difficult it is for anything nonrandom to remain that way over time. Mutation happens.

Even if our genes weren't subject to the unavoidable vagaries of these rare but inevitable copying errors, it is worth emphasizing that our constituent genes (whether mutated or not) most assuredly aren't us. Rather, we are merely temporary combinations of physical stuff, nonrandomly structured so as to produce an identifiable "self" for a very brief time. Compared to our genes, we are temporary indeed, but as already explained, even those potentially immortal genes really don't last forever. *Anitya* underlies all.

In a manuscript titled "Living and Thinking about It: Two Perspectives on Life," the psychologists Daniel Kahneman and Jason Riis make an intriguing distinction between the "remembering self" and the "experiencing self":

An individual's life could be described—at impractical length—as a string of moments. A common estimate is that each of these moments of psychological present may last up to 3 seconds, suggesting that people experience some 20,000 moments in a waking day, and upwards of 500 million moments in a 70-year life. Each moment can be given a rich multidimensional description. An individual with a talent for introspection might be able to specify current goals and ongoing activities, the present state of physical comfort or discomfort, mental content, and many subtle aspects of subjective experience, of which valence is only one. What happens to these moments? The answer is straightforward: with very few exceptions, they simply disappear. The experiencing self that lives each of these moments barely has time to exist.

When we are asked "how good was the vacation," it is not an experiencing self that answers, but a remembering and evaluating self, the self that keeps score and maintains records. Unlike the experiencing self, the remembering self is relatively stable and permanent. It is a basic fact of the human condition that memories are what we get to keep from our experience, and the only perspective that we can adopt as we think about our lives is therefore that of the remembering self. For an example of the biases that result from the dominance of the remembering self, consider a music lover who listens raptly to a long symphony on a disk that is scratched near the end, producing a shocking sound. Such incidents are often described by the statement that the bad ending "ruined the whole experience." But, in fact, the experience was not ruined, only the memory of it. The experience of the symphony was almost entirely good, and the bad end did not undo the pleasure of the preceding half hour. The confusion of experience with memory that makes us believe a past experience can be ruined is a compelling cognitive illusion. The remembering self is sometimes simply wrong.[3]

In addition, isn't it also possible that we are simply wrong in equating the remembering self with the self itself? Although the above was written by Kahneman and Riis in the context of technical assessments of happiness or well-being, it is nonetheless at least as applicable to questions of who or what is our "true self," as well as whether such as self exists at all. Whether or not it does, there is little doubt that it doesn't persist as a static entity. In fact, it doesn't persist at all.

Turning now from the remembered self to the future self, a team of Harvard University psychologists recently reported on what they call the "end of history" illusion, which is the widespread tendency for people to underestimate how much they are going to change in the future:

We measured the personalities, values, and preferences of more than 19,000 people who ranged in age from 18 to 68 and asked them to report how much they had changed in the past decade and/or to predict how much they would change in the next decade. Young people, middle-aged people, and older people all believed they had changed a lot in the past but would change relatively little in the future. People, it seems, regard the present as a watershed moment at which they have finally become the person they will be for the rest of their lives.

The resulting research article is unusual for scientific literature in that it is written in accessible, nontechnical language. Accordingly, here are more of the authors' own words:

> At every stage of life, people make decisions that profoundly influence the lives of the people they will become—and when they finally become those people, they aren't always thrilled about it. Young adults pay to remove the tattoos that teenagers paid to get, middle-aged adults rush to divorce the people whom young adults rushed to marry, and older adults visit health spas to lose what middle-aged adults visited restaurants to gain. Why do people so often make decisions that their future selves regret?
>
> One possibility is that people have a fundamental misconception about their future selves. Time is a powerful force that transforms people's preferences, reshapes their values, and alters their personalities, and we suspect that people generally underestimate the magnitude of those changes. In other words, people may believe that who they are today is pretty much who they will be tomorrow, despite the fact that it isn't who they were yesterday.

After presenting their supporting data (which are impressive, well-documented, and scientifically important, although dispensable for our purposes), the psychologists conclude:

> Both teenagers and grandparents seem to believe that the pace of personal change has slowed to a crawl and that they have recently become the people they will remain. History, it seems, is always ending today.[4]

Earlier, the authors hypothesized that this personalized "end of history" effect likely arises because most people have a comparatively high opinion of themselves, which makes them reluctant to consider that they will change substantially in the future, and/or the widespread difficulty of predicting the future relative to recalling the past. Either way, whatever the cause of the "end of history" illusion, it is, well, illusory. A deep Buddhistic truth underlies these findings: we are always changing, ever impermanent, far more than most of us realize. We cling, nonetheless, to the mirage of our own constancy: believing that although things will change around us, we ourselves will stay pretty much the same. But this is no more feasible or rewarding than trying to grasp an ocean breeze.

There is a famous legend in which a seeker named Hui-k'o had been standing in deep snow seeking an audience with the master Bodhidharma. When finally admitted, Hui-k'o begged to have his agitated self pacified. Bodhidharma replied, "Show me your self and I shall pacify it." "I've sought it for many years, but can't get hold of it," answered Hui-k'o, whereupon Bodhidharma responded, "There, it is pacified once and for all." The idea that all living things—including ourselves—are merely temporary and therefore cannot be grasped (and thereby pacified) may seem a tragedy, a truism, even a platitude, but this makes it no less true; moreover, it merits being pointed out. Most of us—especially in the West—tend to forget or not to realize how deeply *anitya* runs, not just with respect to our personalities and self-image, but at a deeper, material level as well.

Among the molecules of oxygen, nitrogen, and carbon dioxide that we exchange with the atmosphere every time we inhale and exhale, some are the exact same molecules that were inhaled and exhaled by the soldiers at Valley Forge, and even—with a certain statistical probability—by George Washington himself. Some of those molecules also passed through the bodies of a rhinoceros from Africa, an Australian termite, a long-extinct dodo bird from Mauritius, even a *Tyrannosaurus rex*, facts that illuminate *anitya* no less than *anatman*.

For hundreds of millions of years, these and other substances have passed through the living bodies of this planet, thereby connecting them—not just metaphorically but in genuine, tangible reality like the liver and lungs of the same creature, nourished from the same bloodstream. "Just passing through" is what we might say, disparagingly, of a transient journey to a small town somewhere. But this is precisely what we and all other forms of life are and what we are composed of: stuff that just passes through. All of us are eddies in the same stream. Our separate and independent existences are merely figures of speech: easy to recognize, identify, and name, but no more than temporary formations, composed of the same stuff. And while all this may sound more Buddhist than biologist, any scientist who has worked in natural environments will tell you that talk of, say, a snowy egret independent of its watery habitat filled with food, diseases, predators, competitors, mates, nesting grounds, and so on is worse than meaningless—it is downright deceptive.

As human beings, we too are just passing through. We each have somewhat more than two hundred ancestors who were alive, for example, in the year 1700, and literally tens of thousands of potential ancestors from the year 1066. "If we could go back and live again," we read in *The Education of Henry Adams*,

> in all our...arithmetical ancestors of the eleventh century, we should find ourselves doing many surprising things, but among the rest we should certainly be ploughing most of the fields of the Contentin and Calvados; going to mass in every parish church in Normandy; rendering military service to every lord, spiritual or temporal, in all this region; and helping to build the Abbey Church at Mont-Saint-Michel.

No one swims outside the gene pool. What each of us identifies as "our self" is only a temporary collection of genes drawn from a much vaster, shared genome, destined to dissolve back into that gargantuan universal melting pot whose physical substance is shared with all matter, nonliving as well as living. Think of that eddy in a stream, not really existing by itself, independently, but rather a temporary arrangement of "passing-through stuff," given a name for the time being, and therefore sometimes called "oak tree," "snowy egret," or "person." This is not news to the modern biologist, nor to the practicing Buddhist, two seemingly distinct perspectives that originate very differently yet coalesce remarkably in outlook and insight.

Buddhists talk about "mind-streams" coursing through innumerable generations; evolutionary scientists talk about genes doing the same thing. For Buddhists, our physical existence at any given time derives from karma, the accumulated consequences of our past. For biologists, it's a result of accumulated genetic information, a kind of evolutionary karma achieved via the differential reproductive success of certain bodies and their interpolated genes.

Aside from traditional mutations, in which there are changes in the base pairings (adenine, thymine, cytosine, and guanine) that convey originality to DNA molecules, there are also larger-scale mutations, in which chromosomes break and then re-form, sometimes duplicating themselves or getting lost altogether. Even when no apparent mutation can be identified, the phenomenon of hydrogen exchange also occurs, whereby the hydrogen atoms in any DNA molecule (and there are quite a few) are literally switched for other hydrogen atoms surrounding them. It isn't clear whether in such cases the underlying molecule is made differently as a result: the majority of physicists currently seem to believe that once you've seen one hydrogen atom (or ion), you've seen them all. But this might not literally be true; in any event, substituting one atom for another would seem, at least on the theoretical level, to be another way of generating *anitya*.

It is also clear, to anyone with an eye for biogeochemical cycles—which is to say, for the simple fact that change and exchange take place constantly, between our selves and our surroundings, whether those selves are living or dead—that life as well as death is nothing but a constant process whereby our most fundamental component parts are continually being exchanged and transformed. Hence, Thomas Hardy's poem *Transformations* (1915):

Portion of this yew
Is a man my grandsire knew,
Bosomed here at its foot:
This branch may be his wife,
A ruddy human life
Now turned to a green shoot.
These grasses must be made
Of her who often prayed,

Last century, for repose;
And the fair girl long ago
Whom I often tried to know
May be entering this rose.

So, they are not underground,
But as nerves and veins abound
In the growths of upper air,
And they feel the sun and rain,
And the energy again
That made them what they were!

Biologists and Buddhists alike might be inclined to go Hardy one step further and suggest that the man the poet's grandsire knew, his wife, the "fair girl," and everyone else were never actually made "what they were," if this be taken to imply that they were established—even temporarily—with anything approaching physical or biological fixity. They were not only *anatman*, but also *anitya*.

Living things come to exist for purposes of the observer, if nothing else, when their component parts come together in a recognizable way. But immediately, the organism proceeds to change—both to grow and to decay—until eventually it passes away (an expression, by the way, that does double duty; it is not only a euphemism, but also a literally accurate way of describing the process of death and decay, a passing away of those components that previously constituted the organism in question). As a result, life forms are doubly impermanent: first, in that they come and go, arise and depart in their larger, more readily recognizable form as you, me, this dog, that cat, and so forth; and second, that even while existing, all organisms are in a constant state of often imperceptible change.

The story is told about a king who called his wisdom council together and asked for an observation that would always be true, for all living things, at all times. They agreed upon the following: "This too shall pass."

Thomas Jefferson was quite the polymath.[vii] In addition to politics, he made his mark in agriculture, theology, architecture, and even paleontology. But he

---

[vii] It is said that when hosting a dinner attended by all of the then-living U.S. Nobel laureates, President John F. Kennedy noted in his welcoming remarks that "There has never been a greater concentration of intellectual power here at the White House since Thomas Jefferson dined alone."

wasn't always accurate. For example, here is Jefferson reacting to the discovery of fossilized giant ground sloth bones in his native Virginia: "Such is the economy of nature, that no instance can be produced of her having permitted any one race of animals to become extinct."[5] Of course, we know today that Jefferson's reassurance was misguided: many species that existed in the past have gone extinct. In fact, obliteration has evidently been the fate of 99 percent of all species on Earth, testimony to the ubiquity of *anitya*.

But in Jefferson's day, the notion of species extinction was seen as highly threatening, since it suggested that the future of *Homo sapiens* may also be less than secure, as is the ultimate future of any given person. And so, the author of the Declaration of Independence sought to reassure others that our future—like the future of other creatures—is solid and unchanging.

It is indeed challenging to interpret the immensely multifaceted natural world as having been designed with us—human beings—in mind. But Jefferson and his contemporaries were up to it, concluding, for example, that each of the thirty different species of lice that inhabit the feathers of a single species of Amazonian parrot must have been put there, somehow, because of *Homo sapiens*. Our existence was widely thought to be so important that we would never be ignored or abandoned. An accomplished amateur paleontologist, Jefferson remained convinced that there must be mammoths lumbering about somewhere in the unexplored arctic regions; similarly with the giant ground sloths whose bones had been discovered in Virginia, and which caused consternation among biologists as well as theologians of Jefferson's time.

According to Buddhist tradition, Siddhartha Gautama had been living a highly privileged life as a young Hindu prince when he happened to travel outside his protected royal compound and encountered—for the first time in his previously sheltered existence—a person who was very sick, another who was very old, and yet another who was very dead. He was shocked by these discoveries, which induced him to abandon his princely, privileged life and search for wisdom and enlightenment. Eventually he found it and became known as the Buddha.

Like the young Siddhartha in his pre-Buddha phase or Jefferson in the late eighteenth century, most of us are hard-pressed to imagine ourselves as transitional and impermanent, beings who will inevitably experience illness and death and who (unless we die while young) will also grow old. These three experiences—illness, aging, and death—await us, as they constitute the eventual future for all living things. Also noteworthy, and easy to overlook because it is so obvious, is the fact that illness, aging, and death are profoundly biological in themselves, representing transitions that are as natural as they are unavoidable. When biologists consider changes occurring to those who are relatively young, they call it "development." When considering

changes in the second half of life, they speak of "aging." But either way, impermanence is on the march, even though people observing themselves at any given snap-
shot in time often unconsciously imagine that they are fixed and unchanging.

Moreover, along with this perception of the self as solid and permanent, there is
a correspondingly inaccurate impression of the self as central to the universe. The
Buddha was unusual in recognizing that he was ultimately no different from those
sufferers he witnessed and whose plight so upset him, and he may well have been
unusual, also, in recognizing that he was no more special than anyone else, no matter
how privileged his temporary situation.

Almost by definition, we each experience our own private subjectivity, a personal
relationship with the universe, and the universe, it is widely assumed, reciprocates by
treating each of us as especially important, even though there is no evidence supporting this latter assumption and considerable logic urging that it is untrue. Moreover,
even as the illusion of permanence and centrality may be useful, if not necessary,
to normal day-to-day functioning (culminating, at least for the Freudians, in day-
to-day denial of one's eventual death), seductive centrality when not countered by
the Buddhist wisdom of *anitya* is also responsible for a lot of foolishness and even
mischief.

Consider, for example, the strange case of Tycho Brahe, which, on inspection,
turns out to be not so strange after all. An influential Danish star charter of the late
sixteenth century, Tycho Brahe served as mentor to the great German astronomer
and mathematician Johannes Kepler. In his own right, Brahe achieved astounding
accuracy in measuring the positions of planets as well as stars. But Brahe's greatest
contribution (at least for my purpose) was one that he would doubtless prefer to
leave forgotten, because Brahe's Blunder is one of those errors whose very wrongness
can teach us a lot about ourselves, and about the seductiveness of the idea of species-
wide centrality and permanence.

Deep in his heart, Brahe rejected the newly proclaimed Copernican model of the
universe, the heretical system that threatened to wrench the Earth from its privileged position at the center of all creation and relegate it to the status of just one of
the many planets that circle the sun. But he was also a careful and brilliant scientist
whose observations were undeniable, even as they made him uncomfortable: the five
known planets of Brahe's day (Mercury, Venus, Mars, Jupiter, and Saturn) circled the
sun. This much was settled. Copernicus, alas, was right, and nothing could be done
about it. But Tycho Brahe, troubled of spirit yet inventive of mind, came up with
a solution, a kind of strategic intellectual retreat and regrouping. It was ingenious,
allowing him to accept what was irrefutably true while still clinging stubbornly to
what he cherished even more—to what he wanted to be true. And so, like Jefferson,
who sought to reassure those whose confidence was shaken by evidence of species

extinction, and like the pre-enlightened Siddhartha Gautama, who hadn't considered the unavoidable reality of his own eventual decline and death, Brahe proposed that whereas the five planets indeed circled the sun, that same sun and its planetary retinue obediently revolved around an immobile and central Earth.

Brahean solutions are not limited to astronomy. They reveal a widespread human tendency: whenever possible, and however illogical, retain a sense that we are so important, the cosmos must have been structured with us in mind. And of course, the "us" in question must be not only separate from everything else (i.e., not characterized by *anatman*), but also enduring over time (i.e., not characterized by *anitya*).

Some time ago a newspaper article described a most improbable tragedy: a woman driving on an interstate highway had been instantly killed when a jar of grape jelly came crashing through her windshield. It seems that this jar, along with other supplies, had accidentally been left on the wing of a private airplane, which then took off and reached a substantial altitude before the jar slid off. The woman's family may well have wondered about the meaning of her death, just as so many people wonder about the meaning of their lives given that those lives are going, eventually, to end. There must be a reason, they are convinced, for their existence and also for their most intense experiences.

Just as Tycho Brahe struggled to avoid astronomical reality, many people—perhaps most—simply cannot accept this biological truth: they were created by the random union of their father's sperm and their mother's egg, tossed into this world quite by accident, just as someday they will be tossed out of it by a falling jelly jar, or by a delegation of rampaging viruses, or simply by the accumulated wear and tear of a lifetime. It is a difficult message, one that many people inherently resist, since most of us cherish the illusion that we are too important, too distinct, and too central to the world to be no more than drops in a vast ocean. We can't possibly die.[viii]

This shared illusion of unchanging centrality may also explain much resistance to the concept of evolution. Thus, according to Francis Bacon, "Man, if we look to final causes, may be regarded as the centre of the world... for the whole world works together in the service of man.... All things seem to be going about man's business and not their own."[6] Such a perspective, although deluded, is comforting, and not uncommon. Thus, it may be that most of us put emphasis on the wrong word in the phrase "special creation," placing particular stress on *creation*, whereas in fact the key concept, and the one that modern religious fundamentalists find so attractive—verging on essential—is that it is supposed to be *special*. Think of the mythical

---

[viii] The comedian George Burns, who famously kept performing until almost literally the day he died, was asked on his one hundredth birthday whether he worried much about mortality. "Me?" Burns replied. "I can't die. I'm booked!"

beloved grandmother who lined up her grandchildren and hugged every one while whispering privately to each, "*You* are my favorite!"

We long to be the favorite of God or nature, as a species no less than as individuals, and so, not surprisingly, we insist upon our specialness combined with staying who and what we are—if not forever, then at least long enough to provide ourselves with a sense (albeit a misleading one) of permanency. The center of our own subjective universe, we insist on being its objective center as well, an insistence that does not sit well with *anitya*. But the fact that something is intellectually, emotionally, or theologically comforting does not make it true, and *anitya*—although deeply discomforting to many—is much closer to the mark.

In his celebrated and influential book *Natural Theology* (1803), the Reverend William Paley reflected the reassuring alternative to *anitya*, which comports well with Jefferson's meditation on what we now understand to have been extinct species, when he wrote the following about cosmic beneficence, species centrality, and object permanence:

> The hinges in the wings of an earwig, and the joints of its antennae, are as highly wrought, as if the Creator had had nothing else to finish. We see no signs of diminution of care by multiplication of objects, or of distraction of thought by variety. We have no reason to fear, therefore, our being forgotten, or overlooked, or neglected.

Describing the different kinds of causation, Aristotle distinguished, among other things, between "final" and "efficient" causes, the former being the goal or purpose of something, and the latter, the immediate mechanism responsible. The evolutionary biologist Douglas Futuyma has accordingly referred to the "sufficiency of efficient causes."[7] In other words, since Darwin, it is no longer useful to ask, "Why has a particular species been created?" just as it is not scientifically productive to assume that the huge panoply of millions of species—including every obscure soil microorganism and each parasite in every deep-sea fish—exists with regard to and somehow because of human beings. Similarly, it is no longer useful to suppose that any of us is the center of the universe, or that whatever we are at any given moment, we can count on remaining that way forever.

The truth, I submit—and I suspect the Buddha would agree—is more daunting. The natural world evolved and is constantly evolving as a result of mindless, purposeless material events, and human beings are equally without intrinsic meaning or purpose. "We find no vestige of a beginning," wrote the pioneering geologist James Hutton in 1788, "no prospect of an end." This appeared in Hutton's milestone work of geology, titled *Theory of the Earth; or an investigation of the laws observable in the composition, dissolution, and restoration of land upon the Globe.* Not surprisingly, it had a profound effect on the young Charles Darwin. For some, Hutton's realization is bracing; for others, bleak, if not terrifying. In the final chapter of this book, I shall conclude that our unavoidable *anitya* also leads to some interesting consequences when it comes to questions of the meaning of life.

Of course, maybe I am wrong, and Hutton too, and also Darwin, and Copernicus, and the Buddha. Maybe Tycho Brahe was correct all along in his contention that our planet—as well as our lives—is genuinely central to some cosmic design, with our souls (albeit not our physical substance) eternal and unchanging. Many people contend that they have a personal relationship with God; for all I know, maybe God reciprocates, tailoring his grace to every such individual, orchestrating each falling sparrow and granting to every human being precisely the degree of centrality that so many crave. And maybe, even now, in some as yet undiscovered land, there are modern mastodons, joyously cavorting with giant sloths and their ilk, testimony to the unflagging concern of a deity or, at minimum, a natural design that remains devoted to all creatures, keeping them—and especially ourselves—secure, unchanged, and unchanging forever.

Just don't count on it.

The alternative to a world that is eternal and fixed, populated by creatures that are both independent and permanent, is one that is fluid, dynamic, variable, impermanent, and deeply contingent. According to Stephen Batchelor,

> Gautama [i.e., the Buddha] did for the self what Copernicus did for the earth: he put it in its rightful place, despite its continuing to appear just as it did before. Gautama no more rejected the existence of the self than Copernicus rejected the existence of the earth. Instead, rather than regarding it as a fixed, non-contingent point around which everything else turned, he recognized that each self was a fluid, contingent process just like everything else.[8]

There is a big difference between biology oriented toward static things (anatomy, embryology, paleontology, taxonomy), which largely characterized branches of the science in its earlier, descriptive phase, and modern biology, which concerns itself with dynamic processes (neurobiology, development, evolutionary biology, molecular biology). Reality, as Buddhists see it, is similarly a process rather than a static condition. As such, Buddhism promotes what is truly a middle way between two extreme perspectives: eternal, unchanging persistence on the one hand, and ephemerality or nihilism on the other, in which things arise from nothing and then cease existence altogether.

As we saw in the previous chapter, biologists make an important distinction between genotype and phenotype, the former being the total genome of an organism and the latter, the outward manifestation of the creature in question; thus, a given DNA sequence may contain the recipe for a particular protein or enzyme, which eventually gets translated into an aspect of the creature's anatomy, physiology, or behavior. And significantly, as variable as DNA may be—and it is quite variable indeed—phenotypes (how a living thing looks, sounds, smells, feels, etc.) are more variable yet, depending on the environment that the DNA experiences.

Among the Buddha's observations is that our bodies "should be regarded as deriving from former actions [karma] that have been constructed and intended and are now to be experienced."[9] In chapter 6 we'll look more directly at the biology of karma, but for now let's note that all living creatures—human beings included—are temporary assemblages with a history that depends on past actions (*karma*) and that exist only fleetingly at any given time. This is another way of describing not only personal development—that is, the transition from genotype to phenotype—but also the longer-term process of development of a lineage, which is to say, evolution by natural selection. Just as Buddhism calls itself, appropriately, the Middle Way between extreme asceticism and excessive indulgence, as well as between unchanging petrifaction and unfocused nihilism, gene/environment interaction in biology is also a middle way between genetic determinism—in which genes rigidly control the eventual phenotype—and helter-skelter environmental orchestration, in which living things are tabula rasa, blank slates, whose phenotypic outcome is entirely dependent on what each organism encounters, without any consideration of what it ultimately brings to the interaction.

Both are important, such that either extreme is unrealistic to the point of absurdity. It might help to keep in mind the beloved rabbi who once listened, with great sympathy, as a man from his congregation complained bitterly about his wife. A few minutes later, the rabbi responded with equal sympathy to the wife's complaints about the husband, telling each separately and sympathetically, "I understand. You're absolutely right." (In modern terms, he could "feel their pain.") Then, when both husband and wife were gone, the rabbi's wife, who had been listening all the while, berated him: "First you said the man was right, then you said the woman was right. This is impossible. They can't both be right!" The rabbi paused for a moment, and then replied calmly, "Well, *you're* right too."

With only a few exceptions, such as eye color or blood type, genes alone don't determine any phenotypic outcome. Environment counts. And similarly, even the most prescient biologist cannot predict the characteristics of an organism given only information as to the environment in question, deprived of any clue as to whether the creature reacting to that environment is a penguin, a person, or a persimmon.

"Nothing is my own and everything makes me up," writes Robert Aitken, a major figure in contemporary American Zen Buddhism,

> my parents, grandparents, birdsong, the portraits of Rembrandt, the scent of the Puakenikeni,[ix] and the clasp of my friend's hand. Also forming my being are death in the family, the danger of biological holocaust, and the misunderstandings that are inevitable in daily socializing. I am only my setting—interacting with all my past settings—nothing more. The Nirmanakaya is my uniqueness and yours[x] —and the unique formation of each leaf on a single tree. I have almost the same configuration as you but we have minute variations in our genes, experience, and memory to give us mysteriously distinctive individuality. With a perception of this reality, I sense my particular potential and yours—and the precious nature of each being.[10]

Don't misunderstand: Buddhist change and impermanence is not chaotic. Rather, it speaks to a fundamental aspect of life, the fact that unlike, say, crystals or blocks of granite, living things are defined as beings that are transient, temporary, composed of flow-through rather than stuck-in-themselves and stay-the-same.[xi] Evolution itself *is* change, variation in the genetic makeup of a lineage over time. Thus, its two most salient aspects are the connectedness of each lineage (and thus, ultimately, of all life) and modifications in the temporary components of each lineage as time's arrow flies.

Confidence in the "correctness" of evolution is all right so far as it goes, but often it doesn't go far enough. In particular, to see evolution as simply a historical process is to give it too little credit—and similarly, to grant too little credit to its dynamic *anitya*-ness.

Thus, in the public mind, evolution is widely associated with dinosaurs, "cavemen," dusty museum exhibits, and maybe something about embryos having gills and tails. But evolution is much more immediate, and its implications are downright dramatic. It isn't dead, like the dinosaurs. It is alive and kicking, the driving force behind what people are and what they do, no less than where they came from. It is the dynamic stuff of human loves, hates, fears and desires no less than of bones and other fossils. When we really understand evolution as the ongoing process that it is,

---

[ix] A fragrant flower found in Hawaii. Robert Aitken was the founder and for many years the primary force behind the Diamond Sangha, a Zen Buddhist community in Hawai'i.

[x] The physical body of the Buddha.

[xi] Stuck-in-themselves often becomes stuck-*on*-themselves.

then and only then will we really understand human beings in particular as well as life more generally.

To be sure, there are so-called living fossils, organic lineages that have essentially remained unchanged for hundreds of millions of years. These would include such justly renowned creatures as the coelacanths, fleshy-lobed deep-sea fish long believed extinct until specimens were discovered living at great depths in the ocean off Madagascar, or tuataras—peculiar lizards that possess a pineal eye on the top of their heads and are confined to a few islands off the coast of New Zealand—or horseshoe crabs, which aren't crabs as all but oceangoing relatives of so-called sea spiders. In none of these cases has change been altogether absent, although these animals have certainly varied much less over time than have rapidly evolving creatures such as elephants, horses, or *Homo sapiens*.

The key here is whether the surrounding environment has changed significantly. Thus, there is nothing about life itself that makes evolution mandatory. Evolutionary change only occurs in response to changes (often very slow and subtle modifications, not discernible even to the closely observing scientific eye) in the organism's situation and surroundings. But all living things—even coelacanths, tuataras, or horseshoe crabs—will respond to changes in their environments either by evolving or by going extinct.

Of course, over geological time, nearly all organisms have changed, which is to say, they have evolved, leading to the extraordinary panoply of life that characterizes planet Earth. Not only that, but each individual organism also owes its existence—not just its evolutionary existence, its place in the flow of life, but also its individual body in the present moment—to a rather organized, continuous series of changes over time. These changes are not evolution per se, which involves changes in genetic makeup of a lineage over time, but rather changes in the literal composition of each body. Bodies don't evolve in the biological sense (they remain pretty much—although not entirely[xii]—stuck with the genotype they inherit from their parents), but they change, nonetheless.

Beginning as a fertilized egg, a human being eventually comes into existence as that egg divides and divides again, and keeps doing so until it consists of more than a trillion daughter cells, which in turn differentiate into the skin, muscle, bone, blood, nerve, and other tissues that constitute each of us. And to an extent that is only now being appreciated, these cells typically keep dividing and replacing themselves so long as the organism remains alive. For many years it had been thought that brain cells do not repair, regenerate, or replace themselves to any significant degree, just

---

[xii] There are also so-called somatic mutations, which accumulate during the life of an organism independent of those mutations that also take place within the germ cells in the ovary and testes.

as before Dolly it had been thought that once differentiated, body cells could not change into another kind of tissue, not to mention give rise to a whole body composed of all sorts of different cell types.

Now we know differently, although we are still accustomed to thinking that growth ceases at adulthood. Often this is true.[xiii] But development—which is to say, change—occurs throughout life, leading (whether we like it or not) to senescence and eventually death. There really is a new you coming every day! Until there isn't, at which point the "new you" really is new; more on this in the next chapter when we consider interdependence, "inter-being," and the latter's close cousin, the biology of "reincarnation."

Thus far we have observed living things as temporary eddies of matter situated within a constant flux of evolutionary change over time, as well as each being composed during its lifetime of subcomponents (cells) that regularly replace themselves,[xiv] proceeding in the development of each individual from a single fertilized egg to a body consisting of trillions of cells, only a minority of which are actually one's own. From a strictly scientific perspective, therefore, there is every reason for biologists to join with Buddhists in rejecting what the latter call *svahaba*, fixed and unchanging essence. At our deepest, most essential levels, we have no essence.

"Time is the substance I am made of," wrote Jorge Luis Borges. "Time is a river which sweeps me along, but I am the river; it is a tiger that devours me, but I am the tiger; it is a fire than consumes me, but I am the fire."[11]

Similar thinking has undergirded some of the most important advances in evolutionary biology: not merely the idea that life has evolved and is still evolving (which is to say that it is inherently changeable), but also the fundamental fact that all organisms actually change, all the time. It is impossible to overstate the significance of this realization. "The ability to make the switch from essentialist thinking to population thinking," according to the renowned ornithologist and evolutionary biologist Ernst Mayr, "is what made the theory of evolution through natural selection possible.... The genotype is the product of a history that goes back to the origin of life, and thus it incorporates the 'experiences' of all ancestors."[12]

Essentialist thinking is evident in much of the older literature in biology, which viewed each species as consisting of an essential "type" or "norm," à la Plato. Traditional museum collections used to identify a so-called type species for each genus, reflecting the belief that each taxonomic unit contained a perfect, fixed reference point from which others diverged to a greater or lesser extent. Then, under the influence

---

[xiii] Although not always: certain trees as well as many reptiles and fishes keep growing so long as they are alive.

[xiv] And, as we've seen, whose cells are themselves pasted together from other, more primitive cells that alternately consumed and cohabited inside each other.

of researchers such as Mayr, as well as the geneticist Theodosius Dobzhansky, the paleontologist George Gaylord Simpson, and the botanist G. Ledyard Stebbins, the "biological species" concept—one based on a shifting array of constituent populations—gained ascendancy. As a result, species are currently recognized to be natural populations among whom gene exchange readily occurs, but which are relatively isolated, genetically, from other, similar populations. Plato is out; the Buddha is in, right alongside Darwin. Just where the two belong.

4

Connectedness (*Pratitya-Samutpada*)

THERE I WAS, waiting to be issued a wilderness hiking permit at the North Cascades National Park Ranger Station in the small mountain town of Sedro Wooley, Washington, when a message came crackling in via walkie-talkie from a backcountry ranger: "Dead elk by upper Agnes Creek decomposing nicely. Over."

The report was matter-of-fact, yet startling. If you have ever been close to a large, dead, decomposing animal—especially in the summer—you know that "nice" is likely to be the last word that comes to mind. Gross. Disgusting. Nauseating. Loathsome. Certainly not nice.

Yet there was deep wisdom in the ranger's description: "decomposing nicely" is precisely correct. I have no idea if she was aware of having verbalized such an intriguing paradox (the final "Over" was an especially nice touch). But it doesn't really matter. Her description was accurate, not just biologically but also Buddhistically.

That elk, like the rest of us, is on a journey, going back to its primal stuff, with help from various seemingly unpleasant creatures: maggots, bacteria, beetles, earthworms, and perhaps the occasional nibble from a raven or coyote, all under the watchful eye of an ecologically sophisticated National Park ranger. Deposited in the middle of a busy street, a shopping center, or a dining room, a large decomposing mammal would be inconvenient, to say the least, not to mention a public health menace. But out there by the upper drainage of Agnes Creek in the North Cascade Mountains of Washington State, a dead elk is par for the course as it decomposes very nicely indeed, its constituent parts en route back to the ecosystem from which, once upon a time, they came together to form what we label, for convenience, an "elk."

Biologists understand this. So do Buddhists.

And yet, typically, neither realizes that they have company in their awareness, that they are no more alone or disconnected than is that nicely decomposing elk. The same is true for the general public. Insights about decomposition, transience, meaning (and meaninglessness), complexity, self and not-self, even suffering, are not limited to any one perspective, whether modern science or spiritual wisdom. Each illuminates and reinforces the other, sometimes in surprising ways.

It is possible to obtain knowledge of biology by taking an appropriate course or reading suitable books, ideally supplemented with time in a lab as well as in the field. Similarly with Buddhism, although in this case optimum exposure might also include various meditation retreats and possibly a sojourn in a monastery or nunnery. On the other hand, obtaining wisdom, and not just knowledge, requires breadth as well as depth. This book is based on the premise that one needn't be either a Buddhist or a biologist to appreciate the mutual interpenetrations of the two, which complement each other like a pair of powerful searchlights illuminating the same thing from different angles.

Buddhism is "in," and has been for about 2,500 years (though for only a few decades in the West). Ecology, too, is "in," but it has been for considerably less time: for the most part and in the public mind, only about a few decades, since the first Earth Day in April 1970. Their reconciliation is less a matter of resolving an ancient dispute than of two acquaintances realizing that they have been soul mates all along.

It's a communion that is especially pronounced and satisfying when it comes to ecology and Buddhism, a congruence resulting from shared insights into the nature of reality—indistinguishable from the reality of nature—and our place in the whole business. People who follow ecological thinking including some of our most hard-headed scientists may not realize that they are also embracing an ancient spiritual tradition, just as many who espouse Buddhism—succumbing, perhaps, to its chic Hollywood appeal—may not realize that they are also endorsing a worldview with political implications that go beyond bumper stickers and trendy, feel-good support for a "free Tibet."

In this chapter, we shall look especially at ecology as the iconic example of Buddhist biology—or biological Buddhism, take your pick. Ecology is typically defined as the study of the interrelations between organisms and their environment. As such, it closely approaches Buddhism in its sensibilities. Most biologists read-ily acknowledge that living processes are connected; after all, we breathe and eat in order to metabolize, and biogeochemical cycles are fundamental to life (and not merely to ecology courses).

Nonetheless, biology—like most Western science—largely seeks to reduce things to their simplest components. Although such reductionism has generally paid off (witness the deciphering of DNA, advances in neurobiology, etc.), ecologists in particular have also emphasized the stunningly complex reality of organism-environment interconnection as well as the importance of biological communities—which doesn't refer to the human inhabitants of a housing development.

Although community ecology and complicated relationships among a community's living and nonliving components have become crucial aspects of the discipline, recognizing the existence—not to mention the importance—of such interconnectedness nonetheless requires constant struggle and emphasis, probably because the Western mind deals poorly with boundaryless notions. This isn't because Westerners are genetically predisposed to roadblocks that don't exist for our more fortunate Eastern colleagues, but simply because, for reasons that no one seems as yet to have unraveled, the latter's predominant intellectual traditions have accepted and embraced the absence of such boundaries.

In *The Jungle Book* (1894), Rudyard Kipling—despite his "never the twain shall meet" misstatement—captured the power of this recognition in the magical phrase by which the boy Mowgli gained entrance into the life of animals: "We be of one blood, you and I." Being of one blood, and acknowledging it, is also a key Buddhistic concept, reflected as well in the biochemical reality that human beings share more than 99 percent of their genes with each other. At the same time, there is no reason that Mowgli's meet-and-greet should be limited to what transpires between human beings. After all, just as the jungle boy interacted with other creatures—wolves, monkeys, an especially benevolent snake, panther, and bear, as well as a malevolent tiger—everyone's relationship to the rest of the world, living and even nonliving, is equally intense. Thus, we share fully 98 percent of our genes with chimpanzees, and more than 92 percent with mammals generally; modern genetics confirms that we literally are of one blood.

The interpenetration of organism and environment leads both biologists and Buddhists to a more sophisticated—and often paradoxical—rejection of simple cause-and-effect relationships. Thus, the absence of clear-cut boundaries among natural systems, plus the multiplicity of relevant factors, means that no one cause can be singled out as *the* cause—and indeed, the impact of these factors is so multifaceted that no single effect can be recognized either. Systems exist as a whole, not as isolated causative sequences. Are soils the cause or the effect of vegetation? Is the prairie the cause or the effect of grazing mammals? Is the speed of a gazelle the cause or the effect of the speed of a cheetah? Do cells create DNA or does DNA create cells? Chicken or egg? Organism and environment interconnect and interpenetrate to such an extent that neither can truly be labeled a cause or an effect of the other.

It has long been known, for example, that organisms generate environments: beavers create wetlands, ungulates crop grasses and thereby maintain prairies, while lowly worms—as Darwin first demonstrated—are directly responsible for creating rich, loamy soil. On the other hand (or rather, by the same token) it can equally be concluded that environments generate organisms: the ecology of North America's grass prairie was responsible for the existence of bison genes, just as causation proceeds in the other direction, too. Even as ecologists have no doubt that organism and environment are inseparable, ethologists—students of animal behavior—are equally unanimous that it is foolhardy to ask whether behavior is attributable to nature or nurture—that is, to environment or genotype. Such dichotomies are wholly illusory—something that Buddhists would call *maya*.

Consider, for example, a woman who is five feet six inches tall. It is meaningless to ask whether her height is due to her genes or her upbringing: both are clearly involved. But this doesn't mean that we could attribute, say, three feet of her stature to genes and the remaining two feet six inches to her nutrition, health care, exercise history, and so on. Instead, every inch of her—and every bit of her organic substance that exists within every inch—is attributable to the interaction of her genes and her experience. The mutual involvement of genes and environment in creating every organism is mutual indeed, intertwined and interconnected in every way.

Western images are typically linear: a train, a chain, a ladder, a procession of marchers, a highway unrolling before one's speeding car. By contrast, images derived from Indian thought (which gave rise to both Hinduism and Buddhism) are more likely to involve circularity: wheels and cycles, endlessly repeating. Although there is every reason to think that evolution proceeds as an essentially one-way street, Eastern cyclicity is readily discernible not only in ecology—a discipline that is intensely aware of how every key element and molecule relevant to life has its own cycling pattern—but also in the immediacy of cell metabolism, reflected, for example, in the Krebs cycle, or the wheel of ATP, the basic process whereby energy is released for the metabolism of living cells.

Thus far, we have examined two of the primary themes of biological Buddhism or of Buddhist biology (by now it should be clear that the distinction is for the most part arbitrary): not-self or *anatman* and impermanence or *anitya*. In concert, these lead logically and indeed necessarily to interconnection and interpenetration, or—to continue our use of the Sanskrit—*pratitya-samutpada*. The reality of both *anatman* and *anitya* cannot be separated from *pratitya-samutpada*, just as the organism and environment are biologically inseparable. Not-self is genuine and true precisely because each self is, as we have seen, composed entirely of not-self elements, which is another way of saying that all things interconnect and interpenetrate—that is, they all partake of *pratitya-samutpada.*

By the same token, impermanence or *anitya* is genuine and true precisely because all living things are in a constant state of flux, exchanging various parts of "themselves" with their surroundings. We can detect a strong whiff of *anitya* emanating from the ancient Greek philosopher Heraclitus when he announced that "all things change; nothing remains the same," which led, in turn, to his famous observation that one cannot step in the same river twice.[i] Precisely this state of constant flux, of becoming rather than having arrived, is fundamental to—and in many ways, indistinguishable from—*pratitya-samutpada,* which should be distinguished from mere immateriality, inconsequentiality, or some sort of mystical nonsense in which objects in general and organisms in particular are sometimes alleged to be unreal or insubstantial. What they lack is neither reality nor substance, but rather separateness and permanence; in other words, they partake of *anatman* as well as *anitya,* which qualities are key to the fact of their interconnectedness, their *pratitya-samutpada.*

This term has traditionally been translated as "dependent origination" or "codependent arising," which also emphasize the mutual causation and interdependence of all things. It is worth noting, however, that *pratitya-samutpada* has other implications, especially as employed in ancient Buddhist scripture: in its initial appearance, *pratitya-samutpada* provided a way of understanding the psychological origin of *dukkha* (suffering). According to this originalist tradition, *pratitya-samutpada* refers specifically to the so-called twelve *nidanas,* all of which are regrettable: they begin with ignorance, and proceed through "mental formations," consciousness, and such phenomena as feeling, craving, and so on, concluding with birth, aging, and dying, and then over again, indefinitely. These twelve *nidanas* are thus associated with grasping, craving, clinging, and other aspects of enmeshment in the physical world, manifestations of ignorance that must be overcome if one is to arrive at genuine enlightenment.

The process whereby *pratitya-samutpada* has been transformed via Buddhist modernism from a complex trap born of ignorance into a far more positive celebration of beneficent interconnectedness is beyond the scope of this book, although it has been very effectively chronicled by David McMahan in his excellent treatise *The Making of Buddhist Modernism.*[i] McMahan emphasizes that the Buddhist conception of interconnectedness has readily been accepted into modern Western thought in part because the way was prepared by an emphasis on the part of nineteenth-century romantic philosophers and poets on unity and oneness with nature. But

---

[i] To avoid confusion, be aware that this was not intended to be a critique of people's ability to step into a river; rather, it notes that every river is different from one instant to the next.

whereas this European approach was essentially antiscience, the twenty-first-century relevance of Buddhism to modern biology is precisely synchronous with some of the most recent developments in current science.

Even as various scientific and other intellectual disciplines become more specialized (threatening to be composed of people who learn more and more about less and less, until eventually they know everything about nothing), there has been a balancing tendency in the other direction, a perceptible fondness for emphasizing the opposite—interconnectedness, mutuality, and interpenetration—as part of the growing popularity of interdisciplinary approaches generally. Many perspectives can and have issued their own "declarations of interdependence," such that these days scholarly eyebrows would be raised by anyone seeking to make a case for anything's being isolated, unitary, and complete in itself. In this regard Buddhism fits readily into the current cultural moment.

It seems nonetheless true that early Buddhism—as expressed in the most ancient Buddhist texts, the Pali Canon—is arguably quite different from the Buddhism celebrated in this book, whose vision is so similar to modern ecology and other aspects of biology. Thus, the historical Buddha was more concerned with ending human suffering and encouraging individual enlightenment than with promoting environmental awareness and sensitivity. Instead of reveling in connectedness, traditional Buddhist thought focused on the downside of being misled by *maya*, the illusory sense of the material world's importance, such that interpenetration and interdependence are things to overcome rather than to embrace.

Thus, in addition to being psychologized and demythologized, modern Buddhism (especially as promoted and practiced in the West) has also undergone something of an intellectual makeover, placing more emphasis on ecological awareness, social sensitivity, and environmental responsibility than the Buddha or his immediate followers appear to have favored.

A case can indeed be made that the Pali *sutras*, the oldest recorded sayings of the Buddha and thus the ones generally taken as being closest to his original intent, took rather a different view of human connectedness to others, including the natural world. Moreover, this perspective seems to have largely emphasized the *negative* consequences of such connectedness—as noted, dependent origination in its original formulation referred mostly to those aspects of life that bind, enslave, and ultimately disappoint human beings. The foundation story of Buddhism, whereby Siddhartha Gautama achieved enlightenment while sitting under the bodhi tree, revolved around his understanding that the illusory and hurtful aspects of this rootedness in the material world resulted in cravings and other examples of misleading "mental formations."

Even so, there are many striking and subtle connections between Buddhist meta-physics and an ecological stance, notably the recognition that things are in constant flux (*anitya*), which is especially congenial to an evolutionary perspective as well; the recycling of life's fundamental components (associated particularly with *anatman*); and various profound and unexpected consequences of interdependence (*pratitya-samutpada*), as currently understood by most contemporary Buddhists and discussed in the current chapter.

Evolution is the cornerstone of modern biology, leading Theodosius Dobzhansky, one of the twentieth century's greatest geneticists, to famously observe that "Nothing in biology makes sense except in the light of evolution."[2] The deepest take-home message of evolution, in turn, is also the simplest: the fact of connected-ness, of shared history as well as shared current reality. And only Buddhism—not Christianity, Judaism, Islam, Hinduism, or any of the other major world religions—only Buddhism offers a clear statement that this is so, and why. To be sure, there are hints in other traditions, such as Jesus's observation that "Inasmuch as you have done it to one of the least of these, my brothers and sisters, you have done it to me." But in this passage, Christ's concern is clearly his connectedness to other *people*—no trivial matter, to be sure, but much less inclusive than the Buddhist focus on being part and parcel of, inextricably connected to, the least and the greatest of all *things,* living and dead, as well as to everything in between.

At the same time, and as we have noted earlier, there is no single entity labeled "Buddhism," just as there is no single phenomenon identifiable as "Christianity," "Judaism," or "Islam." Certain schools of Buddhism (e.g., Zen) are more sympathetic to ecological ethics than are others (e.g., Theravada, which remains more commit-ted to personal enlightenment). To be sure, the science of ecology is partitioned as well, to some extent between theoreticians (fond of mathematical models) and field-workers (more inclined to get their hands dirty in the real world), but also between ecology as a hard science and ecology in the broader sense of ethical responsibility to a complex natural world. Most spiritual traditions have some sort of moral rela-tionship toward nature built into them, from Christian stewardship to shamanic identification. Yet another reality, and a regrettable one, is that for the Abrahamic religions in particular (Judaism, Christianity, and Islam),[ii] separateness—of soul from body, individuals from each other, heaven from hell, human beings from the rest of the living world, and so forth—is a primary operating assumption.

[ii] People in the West can typically identify exceptions to this unfortunate generalization, instances in which both Judaism and Christianity evince sensitivity to the natural environment rather than their more frequent attitude of separation, control, and human suzerainty. I was particularly pleased, in this respect, to encounter the following, from Islam: "There is not an animal on earth, nor a bird that flies on its wings, but they are com-munities like you" (Qu'ran 6:38).

In a much-loved Buddhist tale, the Chinese sage Fa-tsang is explaining the mutual dependence of all things to the Empress We by pointing to a gold statue of a lion. In his "Essay on the Gold Lion," Fa-tsang argues that

> the arising [of the lion] is due only to dependence, so it is called dependent arising.... The lion is empty [i.e., *anatman*; it is not independently a thing]; there is only the gold...apart from the gold there is no lion to be found.... This means that when we see the lion coming into existence, we are seeing only the gold coming into existence as form.

He goes on, "The eye of the lion is its ear, his ear is its nose, his nose is his tongue, and his tongue is its body. Yet, they all exist freely and easily, not hindering or obstructing each other."[3]

By contrast, dichotomous thinking is especially widespread in the Western worldview, deriving, perhaps, from the Greek Platonic constructs of ideal versus real and intellect versus emotion, and the Judeo-Christian ones of God versus creation, spirit versus flesh, sin versus redemption, and human beings versus nature. Buddhism and ecology, on the other hand, focus on life as an instantaneous aggregate, a complex whole that transcends simple linear, binary distinctions. All things depend, literally, on all other things for their existence: *pratitya-samutpada*.

Ecologists know, for example, that certain flowering plants depend on hummingbirds for pollination and nutrition and vice versa, that the food web binding orca, seal, and salmon does not merely express their trophic relation but defines both the conditions for their survival and their literal moment-to-moment existence. (And this, of course, also speaks to the question of not-self, developed in chapter 2, as well as impermanence, from chapter 3.) It bears pointing out that in such cases, the medium and the message are suitably bound up with each other and reflected in the present book as well: the substance of each chapter flows into that of every other, just as it does in life.

In modern ecology, systems analysts uses complex flowcharts to represent the elaborate relationships of ecosystem components. Although both biology and Buddhism deeply engage the human mind, in both cases there is a shared distrust of simple linear thinking, because the natural world itself is not linear. Not coincidentally, the ideogram that is characteristic of many Eastern languages provides for a unified picture with emancipation from restrictive linear thinking.[iii] Significantly, biology researchers, ranging from ecologists to cancer specialists, are appreciating

---

[iii] Note, by the way, that not all Asian languages employ pictograms; most South Asian languages, as well as Tibetan, do not.

the coordinated complexity of natural communities as well as of disease processes, and the need for the simultaneous apprehension of aggregates.

At the same time, "interaction" includes "action" as well as "inter." The connectedness of events and of matter implies a dynamic state, with components intermingling to varying degrees at different times, constantly changing themselves even as they are changed. This suggests the evolutionary (and Buddhist) concept of life as process, as a constantly shifting nexus of reality spanning a receding past and a progressing future. The dynamics of natural communities dictate moment-by-moment adjustments in interactions between predator and prey, mutualists, symbionts, breather and breathed, shader and shaded, and so on. In addition, such changes are superimposed upon daily cycles, seasonal progressions, serial succession, and ultimately, the ebb and flow of gene frequencies. Cells, too, undergo stages of their own life cycle, as well as a constant and essential barrage of biochemical interconnections with other cells. At every level, things are remarkably complex, and with the growth of modern biology, appreciation of nature's complexity has likewise grown, such that complexity and interconnection are an axiom of today's biology no less than of Buddhism.

Whereas we used to say—to indicate that something wasn't terribly complicated and difficult—that it "isn't rocket science," these days, the standard for complexity is more likely to be ecological communities, neurobiology, or genomics.

Among the most profound of biology's numerous complex interconnections is that between organisms and their environment, notably the understandable yet misleading tendency to consider organism and environment as separate and distinct. The reality is quite different. As the geneticist Richard Lewontin has emphasized, there is a fundamental and irrevocable interdependence between these modes of being, with each defining the other. "The environment is not a structure imposed on living things from the outside," he writes, "but is in fact a creation of those beings. The environment is not an autonomous process but a reflection of the biology of the species. Just as there is no organism without an environment, so there is no environment without an organism."[4]

More recently, Lewontin—a major contributor to the field of population genetics, which has essentially provided much of the mathematical underpinning of modern evolutionary science—goes on to note that the key consideration for evolutionary research is neither organism nor environment, but the interaction between them: "An environment is something that surrounds or encircles, but for there to be a surrounding there must be something at the center to be surrounded. The environment or an organism is the penumbra of external conditions that are relevant to it because it has effective interactions with those aspects of the outer world."[5] Upon occasion, Buddhistic concern can seem a bit overblown—at least to this Western observer—as when the definition of sentient beings is extended to plants or even rivers and mountains. In such cases, however, it is wise to pay special attention to the ways in which such a focus expands our horizons instead of worrying whether it is too expansive. One of the most notable Buddhist writers in this tradition was Dogen, the seventeenth-century founder of the Soto sect of Japanese Zen Buddhism; his best known work, the Mountains and Waters Sutra, includes this gem:

> It is not only that there is water in the world, but there is a world in water. It is not just in water. There is also a world of sentient beings in clouds…in the air.…There is a world of sentient beings in the phenomenal world. There is a world of sentient beings in a blade of grass.…Wherever there is a world of sentient beings, there is a world of Buddha ancestors.[6]

It bears emphasizing that Dogen's extraordinary sensitivity to natural connectedness is not characteristic of Buddhism as a whole. As noted previously, there are elements in Buddhism that focus primarily on individual enlightenment and that consider connectedness a problem rather than a blessing, even as there are others that seem to prefigure the most recent view of ecological systems as changeable, dynamic, and prone to disturbance, yet also highly interconnected and interdependent. And who are we—especially in the West, with its numerous sects of Christianity alone, not to mention the immense disparity between the initial pacifist teachings of Jesus and the historical record of Christianity as one of the great warrior traditions of human-kind—who are we to declare which version of Buddhism is more authentic, legitimate, and worth following?

All religions undergo transitions from their initial formulation to current times. Whether they get better, worse, or remain pretty much the same, however, is subject to dispute. A fundamentalist interpretation of most religious traditions (those that take the Bible or Qu'ran, for example, as literal, incontestable truth) is especially likely to end up in conflict with science rather than in concert. The contrast between fundamentalist and functionalist interpretations can also be seen in ongoing political disputes in the United States with respect to the Constitution, between (mostly conservative) advocates of so-called original intent, some of whom even view that document as having been divinely inspired, and modernists, who maintain that to be relevant in the twenty-first century, the Constitution—like everything else—must be interpreted in a modern context.

To be maximally relevant to today's world, not only political documents but also religious and philosophical traditions are well advised to remain open rather than closed, flexible rather than rigid, connected rather than isolated; that is, to maximize their own degree of *anatman*, *anitya*, and *pratitya-samutpada*. For our purposes, it is both illuminating and useful to take what is currently the most widespread and modern interpretation of *pratitya-samutpada*: interconnectedness and interpenetration as an ecologically sophisticated statement of how the world is, recognizing that even as *pratitya-samutpada* can indeed give rise to *dukkha* (see chapter 5)—it has its beneficial aspects as well.

This perspective has been developed with particular clarity by Thich Nhat Hanh, who has established a new school of Buddhism, which he calls the Tien Hiep Order, the Order of "Interbeing." It appears that Hanh has adopted the phrase *tien hiep* not simply because it is Vietnamese (as he is) and of Chinese origin, deriving from a particular Zen school (*Lin Chi*), but because it enables the knowledgeable Buddhist student to embrace interbeing without the negative associations conjured up among traditionalists by the older, originalist use of the term *pratitya-samutpada*. In this book, I use *pratitya-samutpada,* so as to be consistent with other Sanskrit terms, but note that it is fully interchangeable with *Tien Hiep*. The result is a potent scientific and philosophical tripod…or tricycle…or three-legged stool.[iv] However designated, it is an important foundation of a new and notable aspect of Buddhist practice: engaged Buddhism, whose very engagement derives from an open-minded, open-hearted, expansive understanding of ecology, which embraces not only human beings but all of life.

Earlier we encountered Thich Nhat Hanh's renowned meditation about seeing a cloud in a sheet of paper, which is of course a metaphor for seeing nothing less than everything in everything else. When Thich Nhat Hanh teaches about interbeing, he is touching on *the* fundamental insight of biological Buddhism: the interconnectedness and interdependence of all things. In the Theravada tradition, devotees have long been advised to practice self-restraint and consideration for others; in the other great Buddhist perspective, Mahayana, followers are enjoined to liberate all beings—including but not limited to the human variety—from suffering.

These two major branches of Buddhism differ in significant ways. Theravadan tradition (consistent with Buddhism's ancestral form, which originated in India) is generally more oriented toward asceticism and rejection of the material world; it tends to distrust the senses and sensory experience and is more prone to emphasize the spiritual life rather than immersion in and connection to the natural world. By contrast, Mahayana Buddhists are more inclined to see the universe as an interconnected organism and to emphasize the unity of nature and the positive aspects of natural processes.

[iv] These metaphors are offered in case "trinity" is too evocative of Catholic dogma and "triad" is too reminiscent of a superpower's strategic nuclear arsenal—as both are for me.

Nonetheless, Theravada and Mahayana are sufficiently alike to be labeled "Buddhism," just as both Protestants and Catholics are legitimately called "Christian." Buddhists of every stripe share not only a reverence for the Buddha, but also a deep commitment to compassion, which means something quite different from empathy, sympathy, doing good, being nice, or easy phrases about "feeling your pain."

Modern biologists understand that human nature—our genome—is the accumulated result of organic experience gathered over hundreds of millions of years, during which some manifestations of life (bodies) were more successful than others in projecting aspects of themselves (their genes) into the future. As this process continued, the eventual result was ourselves, along with all other living things. We and our living cousins are thus the result of innumerable prior beings and will in turn be the cause of others in the future. And so, when traditional Tibetan Buddhists repeat, over and over, that all beings have at some time been our mothers, just as we will at some time be theirs—they are altogether correct, not just as mystics but as evolutionary biologists.

Some Buddhists, especially those in the modern tradition of engaged Buddhism, go further and emphasize how the interconnection among people leads to a profound sense of interpersonal responsibility. Thich Nhat Hanh, for example, writes about Vietnamese boat people escaping the aftermath of their terrible war, many of them ravaged by sea pirates, often from Thailand. He was particularly moved by the true story of a twelve-year-old girl, who, having been raped by one of these sea pirates, subsequently threw herself into the sea and drowned. After meditating on this tragedy, Hanh wrote the following poem, titled "Please Call Me by My True Names." It includes the startling suggestion that we ought to feel compassion not only for the girl but also for the sea pirate:

> Don't say that I will depart tomorrow—even today I am still arriving.
> Look deeply: every second I am arriving to be a bud on a spring branch,
> to be a tiny bird, with still-fragile wings, learning to sing in my new nest,
> to be a caterpillar in the heart of a flower, to be a jewel hiding itself in a stone.
> I still arrive, in order to laugh and to cry, to fear and to hope.
> The rhythm of my heart is the birth and death of all that is alive.
> I am the mayfly metamorphosing on the surface of the river.
> And I am the bird that swoops down to swallow the mayfly.
> I am the frog swimming happily in the clear water of a pond.
> And I am the grass snake that silently feeds itself on the frog.
> I am the child in Uganda, all skin and bones, my legs as thin as bamboo sticks.
> And I am the arms merchant, selling deadly weapons to Uganda.
> I am the twelve-year-old girl, refugee on a small boat,

who throws herself into the ocean after being raped by a sea pirate.
And I am the pirate, my heart not yet capable of seeing and loving.
I am a member of the politburo, with plenty of power in my hands.
And I am the man who has to pay his "debt of blood"
to my people dying slowly in a forced-labor camp.
My joy is like spring, so warm it makes flowers bloom all over the Earth.
My pain is like a river of tears, so vast it fills the four oceans.
Please call me by my true names,
so I can hear all my cries and my laughter at once,
so I can see that my joy and pain are one.
Please call me by my true names, so I can wake up,
and so the door of my heart can be left open, the door of compassion.

The touchstone, as we have seen, consists of several fundamental Buddhist teachings that are among the most difficult for Westerners, notably *anatman* (not-self), *anitya* (impermanence), and *pratitya-samutpada* or *tien hiep* (interconnectedness). By this point it should be clear that the Buddhist tripod requires less denial of common sense than one might think; indeed, like most three-legged stools, it's quite stable, constituting a solid foundation for both Buddhism and biology.

One of the fundamental Buddhist teachings, the Avatamsaka Sutra, makes a point very similar to that of interbeing. It speaks to a deep truth: I am because you are, and you are because I am. As Thich Nhat Hanh emphasizes, we "inter-are." Or as the Buddha once taught, in one of his extraordinarily limpid, amazingly simple, yet profound observations: "This is like this because that is like that. And that is like that because this is like this." (One might go further: this is like that because that is like this, and so forth...but you get the point.)

It is widely assumed, especially by believers in special creation, that if there is complexity and order in the natural world—which there certainly is—there must be an overarching force, or Divine Will, that created it. Hence, the so-called argument from design, made famous by the Reverend William Paley, who argued that just as the existence of a watch presupposes a watchmaker, the complexity of living things demands the existence of a divine watchmaker. Such things ostensibly cannot arise from "mere chance"—that is, from evolution by natural selection. But in fact, evolutionary biologists understand that natural selection is essentially the precise opposite of mere randomness, since it operates by small, statistically significant deviations from chance, which as they accumulate over time generate results that are impressively *non*random.

The complex photosynthetic mechanism of green plants, the elaborate biochemistry whereby living cells derive energy from organic molecules, the detailed structure of a vertebrate kidney, the sophisticated neurobiology of *Homo sapiens*—all of these extraordinarily precise natural phenomena derive as a mathematical necessity from the fact of differential reproduction, whereby some genes and gene combinations are more successful than others in projecting copies of themselves into the future; in other words, they are the result of natural selection.

Biologists' focus on evolution occurring at the level of the gene has given rise to criticism from some quarters that excessive reductionism is dangerously afoot. Paradoxically, the debate over reductionism is itself a complex matter. In many ways, science has been most successful when it operates in precisely this way, analyzing the world by examining those component parts that are the smallest. Albert Einstein, after all, once defined science as an effort to explain the greatest number of phenomena using the fewest hypotheses.

At the same time, however, many important scientific advances involve recognizing the relevance of systems and their complexity. Often—though not always—the whole really is more than (or at least, different from) the sum of its parts. A good example is water, which we know is composed of H_2O molecules. But if one wishes to understand hydrodynamics, the "behavior" of water molecules when aggregated is far more relevant than that of individual hydrogen and oxygen atoms. Similarly, if the goal is to study marine biology, the phenomenon of water remains crucial, but once again, the atomic physics of elemental oxygen and hydrogen is going to contribute only a tiny part of the information needed.

Let's take another example. At the level of consciousness and the mind, it is one thing to assert—to be, in fact, quite certain—that mental activities are nothing but the activities of material, electrochemical, and anatomical processes taking place in the brain, but quite another to assume that one can learn everything there is to know in this regard by examining these processes and these alone. The philosopher Daniel Dennett has coined the phrase "greedy reductionism," which occurs when "in their eagerness for a bargain, in their zeal to explain too much too fast, scientists and philosophers...underestimate the complexities, trying to skip whole layers or levels of theory in their rush to fasten everything securely and neatly to the foundation."[7] According to Dennett, radical behaviorism of the sort expounded by B. F. Skinner exemplifies greedy reductionism, as would attempting to understand a baseball game by describing the movements of individual atoms within a sports stadium. And yet, reductionism often pays dividends.

Not only that, but just as greedy reductionism is potentially a real liability, there may be an equally misleading alternative at the other extreme: what we might call "lazy holism," whose practitioners, instead of intoning "*Om mani padme hum*," endlessly repeat, "It's a system, a complex, interconnected system," as an alternative to the hard work of figuring out exactly how such a system actually works.

How far down the reductionist chain should one go? How far up the scale of complex system analysis? In order to understand how the mind works, should one study, for example, the specific properties of molecules, ions, and subatomic particles? Or neural networks, synaptic connections between nerve cells, and electromagnetic pulses across whole brain regions? The point is that sometimes complexity isn't simply a statement that things are complicated and will take time and effort to disentangle; rather, the very complexity of some things is precisely what we would like to understand!

To make a long and wonderful story short, life itself evolved as a result of interconnections; that is, *pratitya-samutpada*. As the Buddha explained, "Because of this, that." Because the ancestors of a given elm tree left more descendants than did others—because of *this* occurrence in evolutionary time—*that* elm tree exists today. Because of *those* great-great-great-grandparents, *these* people exist today, including you and me. *We* are like *this* because *they* were like *that*. And ditto for everyone.

In short, every existence depends on every other; accordingly, there is no such thing as a solitary self because each of us arises in conjunction with others, dependent on and inseparable from those others.

Although for Buddhists and evolutionists alike, living things *are* because of the prior existence of other living things, this dependent connectedness isn't only a matter of the past reaching into our personal present (a kind of vertical integration), but also a statement of our broad, current unity, in which we are also horizontally integrated with everything else. Thus, for Buddhists and ecologists alike, every part of us is destined to be recycled, even as we obtained those parts via the recycling of other, prior lives. When, in the closing image of Gary Larson's hilarious eco-parable *There's a Hair in My Dirt*, a friendly family of worms concludes by looking smilingly up at the reader as though to say, "We'll be seeing you soon," they are factually, ecologically, and Buddhistically right.

In view of our interconnectedness—the extent to which we all "inter-are"—there arises a new, more tolerant, more Buddhistic, and also more ecologically accurate view of many things: worms, bacteria, fungi, the producers as well as the products of decomposition. Here is Thich Nhat Hanh again:

> Defiled or immaculate. Dirty or pure. These are concepts we form in our mind. A beautiful rose we have just cut and placed in our vase is pure. It smells so good, so fresh. A garbage can is the opposite. It smells horrible, and it is filled with rotten things.
>
> But that is only when we look on the surface. If we look deeply, we will see that in just five or six days, the rose will become part of the garbage. We do not need to wait five days to see it. If we just look at the rose, and we look

deeply, we can see it now. And if we look into the garbage can, we see that in a few months its contents can be transformed into lovely vegetables, and even a rose. If you are a good organic gardener, looking at a rose you can see the garbage, and looking at the garbage you can see a rose. Roses and garbage inter-are. Without a rose, we cannot have garbage; and without garbage, we cannot have a rose. They need each other very much. The rose and the garbage are equal. The garbage is just as precious as the rose. If we look deeply at the concepts of defilement and immaculateness, we return to the notion of interbeing.[8]

Since our existence is not a distinct, separable, and private phenomenon, genuine compassion, in the sense of "suffering with," is easy—in fact, unavoidable—insofar as no one is distinct from the recipient of his or her concern. The Avatamsaka (Adorning the Buddha with Flowers) Sutra consists of ten insights into the significance of interpenetration between beings and their environment.

Connectedness to the rest of the natural world is not an exclusively Buddhist insight, nor is it the shared province of Buddhists and professional ecologists alone. Earlier we encountered John Muir's observation that "everything is hitched to everything else in the universe." Here is that notion, in its fuller context, taken from his memoir, *My First Summer in the Sierra*:

> The snow on the high mountains is melting fast, and the streams are singing bankfull, swaying softly through the level meadows and bogs, quivering with sunspangles, swirling in pot-holes, resting in deep pools, leaping, shouting in wild, exulting energy over rough boulder dams, joyful, beautiful in all their forms. No Sierra landscape that I have seen holds anything truly dead or dull, or any trace of what in manufactories is called rubbish or waste; everything is perfectly clean and pure and full of divine lessons.... When we try to pick out anything by itself, we find it hitched to everything else in the universe. One fancies a heart like our own must be beating in every crystal and cell, and we feel like stopping to speak to the plants and animals as friendly fellow-mountaineers.[9]

Although Muir did much to awaken Americans in particular to the beauty as well as the importance of wilderness, he certainly wasn't the first proponent of *pratitya-samutpada* in the Western tradition. It is prominent—although named differently—in the thinking and writing of many (perhaps most) of the romantics, notably William Wordsworth and Samuel Taylor Coleridge. Even earlier, Alexander Pope, in his poem *An Essay on Man* (1734), celebrated connectedness, not only

poetically, but in a manner that could legitimately be described as valid scientifically
and Buddhistically:

> Look round our world; behold the chain of love
> Combining all below and all above.
> See plastic Nature working to this end,
> The single atoms each to other tend,
> Attract, attracted to, the next in place,
> Form'd and impell'd its neighbor to embrace.
> See matter next, with various life endued,
> Press to one centre still, the gen'ral good;
> See dying vegetables life sustain,
> See life dissolving vegetate again.
> All forms that perish other forms supply
> (By turns we catch the vital breath, and die),
> Like bubbles on the sea of Matter borne,
> They rise, they break, and to that sea return.
> Nothing is foreign; parts relate to whole;
> One all-extending, all-preserving, soul
> Connects each being, greatest with the least;
> Made beast in aid of man, and man of beast;
> All serv'd all serving: nothing stands alone;
> The chain holds on, and where it ends unknown.

Later, William Blake longed "to see the world in a grain of sand and a Heaven in
a wild flower." David McMahan points out that following Blake, Samuel Taylor
Coleridge admired "the intuition of things which arises when we possess ourselves as one with the whole," and he sternly criticized as "mere understanding"
the perception that "we think of ourselves as separated beings, and place nature
in antithesis to the mind, as object to subject, thing to thought, death to life."[10]
Part of the strength of the Buddhism-ecology connection is that it unites scientific understanding with spiritual knowing of the sort that Coleridge would
readily espouse.

Later in the nineteenth century, in his poem, *The Mistress of Vision*, Francis
Thompson put it this way: "All things...near and far, hiddenly to each other, connected are, that thou canst not stir a flower without the troubling of a star." He could
as well have written "without the troubling of a garbage dump." For ecologists, no
less than poets or Buddhists, it is *the* basic rule: connectedness, inseparability, food
webs, trophic levels, community interactions, call it what you will.

Nor is connectedness missing from modern literature. Here is Virginia Woolf, in her novel *Mrs. Dalloway* (1925):

> They beckoned: leaves were alive; trees were alive. And the leaves being con-
> nected by millions of fibres with his own body, there on the seat, fanned it
> up and down; when the branch stretched he, too, made that statement. The
> sparrows fluttering, rising, and falling in jagged fountains were all part of the
> pattern....

And in *Vulture*, Robinson Jeffers recounts watching a turkey vulture who had been eyeing the poet as a potential meal. After commenting that the animal's inter-est was premature ("My dear bird, we are wasting time here. / These old bones will still work; they are not for you"), Jeffers shares this biologically Buddhistic conclusion:[v]

> ...But how beautiful he looked...
> That I was sorry to have disappointed him.
> To be eaten by that beak and become part of him...
> What a sublime end of one's body, what an enskyment;
> what a life after death.

For Shakespeare's Hamlet, "the readiness is all," whereas for Buddhists, biologists and Jeffers, it is connectedness. And connectedness counted for Darwin, too.

Thus, at one point in *The Origin of Species* (1859), Darwin speculated playfully that by keeping cats, English spinsters made London a more pleasant and flowerful place. His idea was as follows: cats, as everyone knows, eat mice; mice, as fewer people realize, occasionally destroy the nests of bumblebees, which are typically dug into the ground; and bumblebees, of course, pollinate flowers; so: more cats, fewer mice; fewer mice, more bumblebees; more bumblebees, more flowers. Therefore, more cat lovers = more flowers! No one has ever tested Darwin's proposal. But many other such connections have been elucidated; they are the stuff of ecology, no less than of Buddhism.

Here is just one example reported in the prestigious journal *Science*, of a proj-ect originally conceived to investigate the causes of periodic infestations of gypsy moths, an introduced pest from Europe that, about one year in ten, causes great damage to forests in eastern North America.[11] Researchers suspected there might be a connection between these periodic gypsy moth outbreaks and the abundance of acorns, because of the intervention of white-footed deer mice, common rodents of

---

[v] Or Buddhistically biological; take your pick.

eastern forests. Mouse populations, like those of gypsy moths, tend to fluctuate. The abundance of deer mice seems to be influenced by the availability of acorns; in fact, deer mouse numbers skyrocket following a good acorn crop, which happens about every two to five years. And white-footed deer mice don't eat just acorns. They are also major predators on the pupae of gypsy moths. So it seemed reasonable that in the immediate aftermath of a heavy acorn crop there would be a large mouse population, which, in turn, would keep the gypsy moths in check. Similarly, following poor acorn years, there should be relatively few mice and, therefore, relatively many gypsy moths.

During the summer of a year following a heavy acorn crop, mice were predictably abundant in an upstate New York forest. The ecologists then removed most of the mice from three patches of forest, each measuring about 2.7 hectares (no small task!). They next proceeded to compare the numbers of gypsy moth pupae in these experimental areas with comparable forest plots from which mice had not been removed. Sure enough, fewer mice yielded more moths; forty-five times more, in fact. The next step was to simulate a banner acorn year. With the help of a local Girl Scout troop, they scattered nearly four tons of acorns over the experimentally mouse-depleted plots. Mice quickly repopulated the place, and in fact, mouse numbers went through the roof. Which demonstrated that more acorns indeed means more mice. But the story isn't over yet.

White-footed deer mice are not only consumers of acorns and of gypsy moth larvae; they are also a biological reservoir for the parasitic organism that causes Lyme disease. Infected mice are bitten by tick larvae; these larvae find their way to deer, whereupon they eventually develop into reproducing adult ticks, which become infected when they feed on the mice that have increased because of acorns. This means that the risk of Lyme disease—to people—can be higher in oak forests a few years after a large acorn crop. Thus, the ecologists found that following a good acorn crop, there was, in conjunction with the increased mouse population, an eightfold rise in the number of tick larvae. There were also greater numbers of deer, attracted by the acorns, which brought along their initially average burden of adult ticks, which bred and bestowed their larvae upon the flourishing mice. Mice occupying acorn-enriched plots had about 40 percent more tick larvae than did the denizens of normal (control) forests.

Are there any practical implications to be derived from all this? For one thing, there is a warning to the medical community: It seems likely—but is not yet confirmed—that Lyme disease outbreaks will be especially likely in so-called mast years, when acorn production is unusually high. (Remember, more acorns mean more mice as well as more deer and therefore more ticks.) For another, foresters might be tempted to try inhibiting gypsy moth outbreaks and thus improving lumber yields if they were to distribute additional acorns or some other mouse food. But this might bring about Lyme disease epidemics. (Although more mice mean fewer gypsy moths, they also mean more ticks.) Alternatively, public health officials, wanting to reduce Lyme disease, might look into various ways of chemically suppressing mast (acorn) production. But this might well bring about gypsy moth infestations. Remember, fewer acorns mean fewer mice, which mean more moths.

The take-home message from all this, however, is one not of quietism but of enhanced respect for the deep connectedness of all things. When the noted evolutionary biologist J. B. S. Haldane famously observed that the living world is not only stranger than we imagine but even stranger than we *can* imagine, he was mostly referring to the extraordinary fact that such interconnections as those linking acorns, mice, moths, deer, and Lyme disease—although unknown and unimagined in Haldane's day—are not the exception but the rule. And we, too, are no exception.

When monk asks a master, "How may I enter in the Way?" the master points to a stream and responds, "Do you hear that torrent? There you may enter." Walking in the mountains, a master asks, "Do you smell the flowering laurel?" The monk does. "Then," declares the master, "I have hidden nothing from you."

Enlightened Buddhists see no divide between the human and the natural. By contrast, dichotomous thinking is basic to non-ecological Western thought, deriving, perhaps, from the Greek Platonic constructs of ideal versus real and intellect versus emotion, not to mention the Judeo-Christian: God versus creation, spirit versus flesh, sin versus redemption and—most important for our purposes—organism versus environment and humanity versus nature. Such thinking has long been anathema to Buddhism, as it now is to ecology.

Traditionally, ecology was defined as the study of the interrelations between organisms and their environments, which approaches the Buddhist embrace of interrelation but is still dualistic; significantly, ecologists now modify this definition to emphasize the fundamental identity of subject and surroundings. We cannot separate the bison from the prairie, or the spotted owl from its coniferous forest. Since any such distinction is arbitrary, the ecologist studies the bison-prairie, owl-forest, or egret-marsh unit. Food webs, such as those connecting mouse, acorn, and gypsy moth, do not merely describe who eats whom, but trace the outlines of their very being. The Buddhist suggestion that an organism's skin does not separate it from its environment but joins it could just as well have come from a "master" of physiological ecology.

A seemingly simple message, this, but in such simplicity can be great power and truth. In a famous essay, titled "The Hedgehog and the Fox," the British philosopher Isaiah Berlin used these animals and an observation by the ancient Greek poet Archilochus as a metaphor for various human styles of knowing. "The fox knows many things," Archilochus wrote more than two thousand years ago, "but the hedgehog knows one big thing."[vi] In the present formulation, the wisdom of ecology and Buddhism are both derived from just one thing, but it too is very big.

Here is Thich Nhat Hanh one more time, reflecting upon this very big lesson and how he learned it from an autumn leaf:

One autumn day, I was in a park, absorbed in the contemplation of a very small, beautiful leaf, shaped like a heart. Its color was almost red, and it was barely hanging on the branch, nearly ready to fall down. I spent a long time with it, and I asked the leaf a number of questions. I found out that the leaf had been a mother to the tree. Usually we think that the tree is the mother and the leaves are just the children but as I looked at the leaf I saw that the leaf is also mother to the tree. The sap that the roots take up is only water and minerals, not sufficient to nourish the tree. So the tree distributes that sap to the leaves, and the leaves transform it with the help of the sun and the air, and send it back to the tree for nourishment. This communication between leaf and tree is easy to see because the leaf is connected to the tree by a stem.

We do not have a stem linking us to our mother anymore, but when we were in her womb we had a long stem, an umbilical cord. The oxygen and the nourishment we needed came to us through that stem. But on the day we were born, it was cut off, and we had the illusion that we became independent. That was not true. We continue to rely on our mother for a very long time, and we have many other mothers as well. The earth is our mother. We have a great many stems linking us to our Mother Earth. There are stems linking us to the clouds: if there are no clouds, there will be no water for us to drink. We are made of over seventy percent water, and the stem between the cloud and us is really there.... There are hundreds and thousands of stems linking us to everything in the universe, supporting us and making it possible for us to be. Do you see the link between you and me?... I bowed my head knowing I have a lot to learn from that leaf.[12]

With dualism overcome and the world seen in its organic wholeness, it is absurd to consider natural processes as harmful to themselves, just as the death of a leaf doesn't hurt the tree; quite the opposite. There is instead a kind of integrity about the natural world: "A duck's legs, though short, cannot be lengthened without dismay to the duck, and a crane's legs, though long, cannot be shortened without misery to the crane."[13] Western thinking has generally been more Procrustean, seeking to amputate, stretch, or otherwise deform the natural world to suit our desires. And yet, ecologists have come to a more Buddhistic realization, with their (admittedly belated) acceptance of processes that are typically seen as negative, although natural—decomposition, predation, even forest fires—as having a place in maintaining healthy ecosystems.

Perhaps one day we will all celebrate Interdependence Day. Or better yet, recognize that it occurs all day, every day.

We have noted that even simple cause-and-effect has been increasingly reevaluated in ecological thinking, with attention to systems analysis and complex flowcharts representing numerous inputs and a bewildering array of interconnections, a kind of neural net writ large. "When this exists, that comes to be," we read in the Assutava Sutra. "With the arising of this, that arises. When this does not exist, that does not come to be. With the cessation of this, that ceases." Not surprisingly, simple verbal description, with its unavoidable linearity, is inadequate for ecologists, just as it has long been disdained by Buddhist masters who are more likely to identify with the image of Indra's net, a structure (and/or a metaphor) described in the Avatamsaka Sutra.

It turns out that Indra, the Vedic emperor of the gods, had a net woven for reasons that are unclear, one suggestion being that it was for the amusement and entertainment of his celestial daughter. In any event, this net is infinitely large, containing a pure and perfect jewel at each of the filamentous intersections. If you look closely at any one of these jewels, you can see reflected therein all the other jewels (remember, the net is infinite). It's a bit like Wordsworth's seeing the world in a flower. Also, remember that on the surface of each of the jewels, all the others are reflected, so when you peer into any one, you not only see all the others, but each of those others reflects all of the remaining others, in turn.

A key Buddhist concept is that all things are reflected in Indra's net, human beings included. Most people in the West are familiar with the saying that "there is no such thing as a single ant." Less well-known is that especially in the East, there is also no such thing as a single person. The cultural and social traditions of much of Asia differ considerably from those of the West, notably when it comes to identifying (or "situating," as humanist scholars like to put it) a person's place in society. "In the West," according to the Buddhist monk Victor Sogen Hori,

The person is an independent being, who exists autonomous from social roles and relations. For most societies outside the influence of the European

Enlightenment, however, a person is not an independent being...quite the opposite, a person has identity and uniqueness only because of his or her social relationships. In answer to the question, "Who are you?" one does not answer with just one's name, but rather, "I am the son or daughter (gender makes a difference) of so-and-so, father or mother to so-and-so, husband or wife to so-and-so, member of such-and-such, resident of such-and-such, etc." My identity as a person depends on my relationships with other persons and, ultimately, with place, land and nation, with history and time. Rights and responsibilities accrue to me in virtue of the social roles and relationships in which I am involved. Not merely personal identity, but also my very nature as a human being is dependent and social.[14]

For a similar perspective, this time emphasizing the deeper reality of how *pratitya-samutpada* defines and delineates our very existence, beyond how it colors social experience in one society versus another, consider this observation by Stephen Batchelor:

The "primal experience" of recognizing our involvement with others reveals that the presence of others is not incidental but essential to our being. We do not just happen to "bump into" others, but are inescapably together with them in the world. As this awareness dawns upon us, self-concern is seen to be a drifting, alien body, untethered from its moorings, that forms no part of who we really are. It is dislodged from its position as the seemingly indestructible center of motivation and turns out to lack any...foundation. Instead of representing an essential element of our being, [self-concern] in fact *conceals* an essential element of our being.[15]

It is worth emphasizing that this involvement with others is a "primal experience" indeed, and that while it includes human beings, it is assuredly not limited to *Homo sapiens*. In fact, many species—perhaps most, if not all—that used to be labeled "solitary" are now known to show some form of sociality, that is, to be involved with other members of their species beyond the minimum necessary for sexual reproduction. For example, research that I conducted nearly four decades ago involved live-trapping adult foxes and raccoons—animals traditionally thought to live solitary lives—and then introducing foxes to other foxes and raccoons to other raccoons, comparing how neighbors interacted with each other versus pairs that had been trapped far apart and who therefore had no opportunity to get to know each other. The results, published in the journal *Science* under the title "Neighbor Recognition among 'Solitary' Carnivores," showed that these supposedly asocial species were in fact quite aware of their neighbors, having formed at least a rudimentary kind of social network.[16]

Nor is social networking limited to vertebrates. One of the most exciting developments in biological science involves the discovery that in many cases, microbes not only are crucial to the lives of their ostensibly independent "hosts" but are significantly more involved with other microorganisms than had previously been imagined. Moreover, quite a number engage in what is now called "quorum sensing," in which individual bacteria emit a chemical that is detected by other cells of the same species, so that when a sufficient amount of the relevant chemical is detected, the bacteria "know" that there are enough of them—a "quorum" is present—for the members to proceed with some important activity.

Quorum sensing was first observed in the bioluminescent bacterium technically known as *Vibrio fischeri*. When these creatures are free-floating in the ocean, they are dark, although each regularly emits a small amount of a chemical to which other *V. fischeri* are sensitive. When a sufficient number of these bacteria are concentrated near each other, as happens when they aggregate in the light-producing organ of a particular species of deep-sea squid, the concentration of their signaling molecule then becomes high enough that the bacteria begin to luminesce, and the squid literally commences to glow in the dark. It would be ineffective for these bacteria to glow when solitary, because the amount of light produced by just one bacterium is insufficient to be seen and their biochemical energy would be wasted; by quorum sensing, these creatures are able to save their energy for those times when there are enough of them to have an impact.

Here's another example, once again from bacteria. It turns out that one reason *Salmonella* can cause such serious disease symptoms ("food poisoning") in people is that once inside a human intestine, the invading bacteria divide up their roles. Some of them invade the intestinal epithelium and generate an inflammatory response in the host/victim. These invading *Salmonella* bacteria do not survive long and are essentially on a kamikaze mission. The body's response to their presence—inflammation—involves, among other things, generating chemicals that react with a sulfur compound already present in the central lumen of the intestine, producing an organic compound known as tetrathionate. And those *Salmonella* still in the gut lumen—unlike the bacteria that normally inhabit a healthy intestine (e.g., *E. coli*)—are uniquely able to metabolize tetrathionate, which gives them a big advantage over the remaining "good" bacteria. Tetrathionate, in turn, provides nutrition for the remaining *Salmonella*, those that hadn't sacrificed themselves to produce that inflammatory response. The end result is that by cooperating with each other, *Salmonella* are able to achieve much more than if they were acting independently,[17] to their benefit—albeit not to ours.

Thus, when Batchelor observed that *pratitya-samutpada* is an "essential element of our being," he was being too restrictive. It's an apt description for living things generally, even some of the simplest organisms. In short, there is an appropriateness to nature and, accordingly, a place for all natural phenomena, including predation, parasitism, forest fires, decomposers, and so forth, in the thinking of today's "skin-out" biologists, notably ecologists and evolutionists, although it is notable how long it has taken for some of these insights to become widespread. Western science has something of a checkered history in this regard, including a decades-long tradition within wildlife management that supported the extermination of predators so as to "improve" deer and elk populations, as well as scientific (actually, *un*scientific) resistance to accepting the appropriateness of natural wildfires.

Gradually, however, the Western stance is becoming congruent with its Eastern antecedents not only among ecologists, but even in the work of medical science; for example, the now-vigorous field of Darwinian medicine emphasizes that many bodily responses (such as fever) long seen as pathological actually reflect a deep, natural "wisdom of the body" whereby elevated body temperature inhibits the reproduction and metabolism of numerous pathogens and promotes the action of certain antibodies. Another practical example from medicine: it has only recently been recognized that it is possible to overprotect infants and children by limiting their exposure to bacteria and antigens, as a result of which they are deprived of the opportunity to develop their own healthy internal array of natural immunities.

In short, *pratitya-samutpada* extends not only through evolutionary time but also across ecosystems and deep into our selves.

Then there is *dukkha*, the existence of which constitutes the first of Buddhism's Four Noble Truths, and which we shall visit in more detail in chapter 6. (The second Noble Truth identifies the source of *dukkha* in craving, the third recognizes that it can be transformed, and the fourth constitutes the Eightfold Path to how this can be accomplished.) It is no coincidence that Henry David Thoreau, America's first great environmentalist, was also a student of Indian religion and the first translator of the Lotus Sutra into English; in this classic teaching, Shakyamuni Buddha compares the *dharma*—the true nature of reality—to a soothing rain than nourishes all beings. And who would dispute that all of us—other beings along with the human sort—could use both nourishment and soothing?

The Buddha urged his followers to be sensitive to the suffering of all sentient beings; thus, his First Precept is to commit oneself to *ahimsa*, or nonharming. In the Mahayana school of Buddhism, the ideal is to go further and to become a *bodhisattva*, an especially compassionate individual who vows to relieve the suffering of all beings rather than merely pursue his own path. In the Metta Sutra, Theravada monks and lay adherents vow to practice loving-kindness: "Even as a mother protects with her life her child, her only child, so with a boundless heart should one cherish all living beings." And here is the first verse of "The Bodhisattva Path," by Santideva, a revered Buddhist poet of the eighth century: "May I be the doctor and the medicine / And may I be the nurse / For all sick beings in the world / Until everyone is healed."

Ecology is many things: a science, a worldview, a cautionary tale. It can be nearly incomprehensible in its mathematical modeling, downright tedious in its verbal pomposity, theoretically abstruse yet dirty-under-the-fingernails practical, often ignored and derided although desperately needed as a voice for basic planetary hygiene and a practical corrective to human hubris. It has been called the "subversive science," since it undermines our egocentric insistence of separateness and with it our inclination to ride roughshod over the rest of the natural world. Buddhism is no less subversive, its ecological implications in particular carrying the serious practitioner far beyond giddy adoration of the Dalai Lama or longing to share a passion with Richard Gere.

To E. M. Forster's celebrated injunction, "Only connect," Buddhists and ecologists would add that we are already connected. The joke goes like this: what did the Buddha say to the hot-dog vendor? "Make me one with everything." (No joke, however: we're already One.)

Our job is to recognize this union and act accordingly, paying attention not merely to our breath but also to our breadth. And depth. However, the ecological implications of Buddhism—or the Buddhist implications of ecology—are neither easy nor for the fainthearted. Shortly after the First National Economic Development Plan was drafted in Thailand, the Bangkok government imprisoned monks for teaching *santutthi* (contentment with what one has) out of fear that this Buddhist ideal would interfere with shortsighted economic development—which it would, and did.

I am aware, however, that environmental policies may often seem greener on the other side of international borders, and I don't want to give the impression that Buddhist cultures are, by virtue of their traditions, more environmentally sensitive than in the West. Regrettably, those few governments worldwide that are primarily influenced by Buddhist traditions (e.g., Sri Lanka, Thailand, Laos, Cambodia, Vietnam, and, to a lesser extent, South Korea and Japan) are not necessarily on the side of the proenvironmental angels.

An anonymous reviewer of an early draft of the present book wrote that

insofar as Buddhism institutions in Asia become enmeshed in protecting and legitimizing the political and social order...they succumb to the same

problems as other "established" traditions: they typically justify political violence, hierarchical social orders, and conformity to conservative social mores,
and are often more interested in protecting their own institutional concerns
than in leading beings to awakening. We don't need to whitewash Buddhist
history in order to find worthwhile ideas and practices in their traditions.

In the United States, thousands of moviegoers chuckled indulgently at a notable
scene in *Seven Years in Tibet* in which Buddhist monks carefully and by hand
removed earthworms from a construction site. But when it comes to certain spiri-
tual-religious-philosophical traditions, beware what you espouse: they may turn out
unexpectedly demanding. Thus, in another movie, *Becket*, we witness the transformation of Richard Burton, as Thomas à Becket, who upon being made Archbishop
of Canterbury gives up not only his sybaritic lifestyle but eventually his life. When
it comes to the interpenetration of Buddhism and ecological wisdom, the consequences of serious "practice" may be less dire, although equally significant: instead
of losing a life, we just might expand ours.

L et us therefore return to that elk, decomposing so nicely, playing out a crucial
principle of both biology and Eastern wisdom: namely, the wondrous banality
of connectedness. The *Katha Upanishad*, ancestral to Buddhism, gives a hint of how
Eastern sages perceived this same connectedness and derived considerable solace
from it a few thousand years ago:

> Finer than the fine, greater than the great, the Self hides in the secret heart of
> the creature....Seated, he journeys far off; lying down, he goes everywhere.
> Realizing the bodiless in bodies, the established in things unsettled, the great
> and omnipresent Self, the wise and steadfast soul grieves no longer.

To be sure, death is often an immediate tragedy, but Buddhism and biology nonetheless coalesce in this wisdom: death is a natural and necessary expression of our
connection to everything else, of our being alive in the first place.

Eventually, our own particular eddy dries up and runs out of stream, and the stuff
of which we are composed begins a more one-directional process. We decompose,
eventually to be recomposed in other ways, materially—albeit not, in my opinion,
spiritually—reincarnated as other creatures. And these, in turn, are merely the latest combinations of the same old stuff. Chips off the good old material block. It is
a good-enough goal for us all: to decompose nicely, starting right now, in a sense,

while still alive. Another such goal is to view our fellow creatures—elk, redwood tree, or human being—as fitting in, belonging, being a good, acceptable, and appreciated part of the world we all share, and to which all will return. According to Edward Abbey, such an attitude is "an expression of loyalty to the earth, the earth which bore us and sustains us, the only home we shall ever know, the only paradise we ever need—if only we had the eyes to see."[18]

It will nonetheless be a challenge, this business of seeing ourselves as united with the Earth and especially our fellow creatures in ways that appear counterintuitive, or—even more challenging—to recognize that we don't exist independently at all. It takes some doing—or rethinking—to live happily with dead elk in our favorite alpine meadows, crabgrass in our lawns, a galaxy of bacteria in our intestines. But when we think of these things as part of a bigger process—a vast natural system that incorporates our tiny, tidy little self-identified lives but is far more expansive, indeed, that is wildly inclusive in the very broadest sense—we can start seeing the wisdom of modern ecology no less than of ancient Buddhism. We can then begin to appreciate Darwin's meaning when he wrote, in the justly famous final paragraph of *The Origin of Species*:

It is interesting to contemplate a tangled bank, clothed with many plants of many kinds, with birds singing on the bushes, with various insects flitting about, and with worms crawling through the damp earth, and to reflect that these elaborately constructed forms, so different from each other, and dependent upon each other in so complex a manner, have all been produced by laws acting around us.

And we might especially agree with Darwin's conclusion, that

there is grandeur in this view of life, with its several powers, having been originally breathed...into a few forms or into one; and that, whilst this planet has gone cycling on according to the fixed law of gravity, from so simple a beginning endless forms most beautiful and most wonderful have been, and are being evolved.

We could even see that there is something suitable, appropriate—downright nice—about being genuine, bona fide parts of our world, the only one we have ever had, and one whose liveliness is made comprehensible not only by modern biology but also by the ancient wisdom of the East.

5

Engagement, Part 1 (*Dukkha*)

IN A NOW-CLASSIC manuscript published nearly half a century ago in the journal *Science*, the historian Lynn White argued that "The Historical Roots of Our Ecological Crisis" derive from a Western religious tradition that itself dates from more than thirty-five centuries ago.[1] That tradition, initially espoused by a largely nomadic, mostly illiterate early Bronze Age tribe of desert dwellers, not only separated humanity—that is, themselves—from the rest of the natural world, but also claimed Old Testament sanction for the view that nature exists *for* them (which is to say, us) and, moreover, that it is therefore our God-given right—in fact, our obligation—to exploit it, even to the point of outright abuse.

Here is Genesis 1:28, from the King James Version of the Bible: "And God blessed them, and God said unto them, Be fruitful, and multiply, and replenish the earth and subdue it; and have dominion over the fish of the sea, and over the fowl of the air, and over every living thing that moveth upon the earth." This extraordinary societal and theological hubris—all that subduing and having of dominion—is not only a personal, biological, and Buddhistic absurdity, it is as persistent as it has been downright destructive.

In this regard, we might take at least some comfort from the several ecumenical movements in the West that have begun to espouse "faith-based stewardship," intended to counter that troublesome Abrahamic theology of human centrality. The idea, in brief, is that human beings have a responsibility to care for God's creation, such that "dominion" includes protective responsibility. But even as we might applaud this development, it is difficult not to register a small shudder of distrust, because even so laudable an enterprise as human stewardship still revolves around the stubborn, persistent idea that We Are Special.[i]

In a sense, there isn't all that much difference between claiming on the one hand that nature exists *for* human beings to exploit, and, on the other, urging that it exists *for* us to protect. Either way, *Homo sapiens* is presumed to occupy a uniquely privileged place in the cosmic scheme, one that distinguishes us from everything else. Even an ethic of stewardship takes it for granted that we and the natural world are separate and distinct, and also that we were created for a purpose, part of which happens to involve taking care of nature—of something external to us.[ii]

Better, of course, to take care of nature than to exploit it, but as Buddhists are likely to conclude, nature is quite capable of taking care of itself—except, perhaps, when people insist on messing it up. And any fair, open-minded look at the world we inhabit—and of which we are an integral part, just as it is a part of us—must conclude that we have messed it up quite badly. There is also no avoiding the fact that human actions have done and continue to do a great deal of harm to other human beings as well—not simply insofar as *Homo sapiens* is part of the greater worldwide ecosystem, but as a consequence of how our actions directly ramify into human social systems.

The First Precept of Buddhism is *ahimsa* ("Do no harm"). Like many such precepts, it is lovely in theory, yet impossible in practice: even if one elects to be vegetarian—and not all Buddhists do—no one can survive without doing harm to carrots, broccoli, rice grains, and other plant foods. Buddhism's First Precept therefore shares something with G. K. Chesterton's famous observation about Christianity—namely, that it has not been tried and found wanting, but rather that it has been found difficult and left untried. It is not just difficult but literally impossible for people to live without inflicting some sort of harm upon other living beings.

To take an extreme case, strict Jains insist that when walking along a sidewalk they must be preceded by sweepers whose job is to brush away any tiny, unseen organisms, lest they be stepped upon.[iii] Such doctrine strikes most of us as ridiculous. Nonetheless, it is not only possible but also desirable—if not essential—to live in a way that minimizes unnecessary harm, a route described in Buddhism's Eightfold Path and, in modernized form, in Thich Nhat Hanh's Fourteen Precepts of the Tien Hiep Order of "Interbeing." Just as the hurtful approach of Judeo-Christian

---

[i] Of course, we *are* special, not only in the sense that every species is, but also because we are able—among other things—to pride ourselves in our specialness. But that doesn't mean that we are outside the ambit of biology.

[ii] It should be mentioned that the Christian stewardship movement is, unfortunately, very much a minority voice.

[iii] The Jains are followers of an ancient religion that originated in India about a century before Buddhism and that especially values nonviolence.

doctrine toward the natural world can be seen as emerging largely from Genesis, the Buddhist-promoted desirability of a thoughtful, protective, and supportive attitude toward our planet generally and toward "all sentient beings" in particular derives from *anatman*, *anitya*, and *pratitya-samutpada*, especially when it comes to modern engaged Buddhism.

Ecological insight leads similarly to prescriptions for environmentally sensitive policies, including preservation of wilderness, sustainable resource utilization, maintenance of biodiversity, reduction of pollution, and so forth. More recently, the prospect (some would say the danger) of cloning, stem-cell research, cross-species gene transpositions, and the like has greatly raised the public and professional profile of bioethics when applied to advances in molecular biology; here, too, as the opportunity arises for doing harm—or good—so does the importance of *ahimsa* and of the benevolent mindfulness that Buddhism espouses.

Ahimsa and its biobehavioral counterparts, responsible stewardship of the environment and mindful awareness of our own actions, lead ineluctably to the question of social and political engagement and the recognition that whatever we do brings repercussions, not just in the world acted upon but in the subsequent life of the actor. In other words: karma.

Buddhism is widely seen as quietistic, self-involved, and concerned only with personal enlightenment and the achievement of inner peace. As we have seen, there is some truth to this characterization (especially when applied to the earliest manifestation of Buddhism), just as some ecologists and evolutionary and molecular biologists are concerned only with their science, refusing to become involved in messy questions of politics, policy, or preservation. However, engaged Buddhism has grown dramatically, especially since the Vietnam War, just as environmental scientists have become more politically active, notably since the first Earth Day in 1970. Serious questions remain to be resolved, nonetheless, such as precisely how one can live in a socially and environmentally responsible manner. At minimum, engaged Buddhism and politically relevant biology both commend the practices of simplicity and self-restraint because of their beneficial effect on our shared environment.

But what, specifically, does this have to do with Buddhism? Historically, perhaps, not very much. As we've already noted, a strong case can be made that the Buddha and his original followers and early intellectual and spiritual descendants were primarily concerned with individual enlightenment and helping people transcend their personal *dukkha*. At the same time, there is a powerful trend within what has been called Buddhist modernism that takes *anatman*, *anitya*, and *pratitya-samutpada* and derives from them a potent recipe for engagement in the world, especially by combining them with sensitivity to *dukkha* as well as responsibility for our actions, which is to say, karma.

Let's take these one at a time, beginning with *dukkha*.

For Buddhist thought, suffering isn't something imposed upon human beings because they have sinned or otherwise displeased God. Indeed, Buddhists disavow any belief in an omnipotent creator deity, whether singular or plural. Suffering is part of the nature of the world, a fact of existence.[iv] There is no need, accordingly, for any elaborate Buddhist theodicy (the effort to justify one's belief in an omnipotent and omnibenevolent deity in the light of so much pain and suffering in the world). As we'll see, for many Buddhists—especially those following the earliest traditions—karma might suffice, since one can always argue that suffering is the result of a karmic burden acquired during one or more previous existences. It is a dispiriting message, however, to claim that when an infant dies of malaria, for example, she is getting what she deserves because of prior misdeeds.

For most Westerners, at least, a far more reasonable interpretation is that there are certain "built-in" sources of suffering, which essentially come with the territory as direct consequences of our existence in the world. The primary foundation story in Buddhism tells how the young Siddhartha Gautama grew up as a pampered prince around 500 BCE, sheltered from all possible sources of sadness and unhappiness. Then, at the age of twenty-nine, he sneaked out of his castle. He was shocked at what he saw: a very old man, someone who was very sick, and a corpse. Appalled by these indicators of *dukkha*—of the misery that awaits everyone—the prince left his luxuries, as well as his family, in search of enlightenment, which he eventually obtained while meditating under the bodhi tree.

The story has it that the Buddha went through a long period of self-denial, mortifying his body in his desperate search. Supposedly (and nonsensically), he ate less and less until his food intake was down to one grain of rice per day. Eventually, sitting under that tree, the—presumably very skinny—Siddhartha Gautama encountered a young girl who offered him some rice milk, which he accepted. This acceptance is considered significant, an acknowledgment that we need body as well as spirit, that the body counts, and not simply as offensive flesh to be mortified. The Buddha's consumption of rice milk is thus emblematic of the Middle Way, a denial of dualism, announcing that correct living is not a question of mind *or* body, good *or* evil, all *or* nothing, self versus the rest of the world.

To some extent, the term "engaged Buddhism" is an oxymoron, along with "environmentally sensitive Christianity," since each requires a departure from its earliest traditions. In the case of Buddhism, however, the stretch isn't nearly as great, since, when the Buddha agreed to take material sustenance, he acknowledged the validity

---

iv This is the first of Buddhism's Four Noble Truths, the fact that *dukkha* is not only universal but also, to a large extent, unavoidable.

and importance of regular, run-of-the-mill stuff. Moreover, in its earliest texts (the Pali Canon) Buddhist doctrine includes numerous admonitions that followers should treat the natural world—plants, animals, rivers, mountains, even deserts and rocks—with respect and even love, that is, with *ahimsa*.

Here is a word likely to be new to most readers: soteriology. It refers to issues associated with spiritual redemption, especially as employed by Christians concerned about their personal salvation. The relation to Buddhism? Simply this: as already noted, there is reason to think that the earliest forms of Buddhism focused more on soteriology than on social and environmental responsibility. But this soteriological concern is being increasingly supplanted by a wider focus, one that makes sense not only ethically but also biologically. The present book, in fact, is concerned not only with identifying parallels and convergences between Buddhism and biology (its avowed purpose), but also with promoting this wider focus, aligning itself with a gently subversive movement within mainstream Buddhism that Stephen Batchelor calls "Buddhism 2.0."

"To use an analogy from the world of computing," he writes,

> the traditional forms of Buddhism are like software programs that run on the same operating system. Despite their apparent differences, Theravada, Zen, Shin, Nichiren, and Tibetan Buddhism share the same underlying soteriology, that of ancient India outlined above. These diverse forms of Buddhism are like "programs" (e.g. word processing, spreadsheets, Photoshop etc.) that run on an "operating system" (a soteriology), which I will call "Buddhism 1.0." At first sight, it would seem that the challenge facing the dharma as it enters modernity would be to write another software program…that would modify a traditional form of Buddhism in order to address more adequately the needs of contemporary practitioners. However, the cultural divide that separates traditional Buddhism from modernity is so great that this may not be enough. It might well be necessary to rewrite the operating system itself, resulting in what we could call "Buddhism 2.0."[2]

For any computer mavens reading this book who also happen to be Buddhist traditionalists, I'd like to add this reassurance: "Buddhism 2.0," as envisioned by Batchelor, represents a software update; it does not require installing any new chips or a new hard drive. It is still Buddhism.

Meanwhile, back in the formative world of Buddhism 1.0, as a result of his insights, young Gautama became known as the Buddha, whereupon he spread his newfound wisdom to increasing numbers of eager disciples. Bear in mind that the Buddha's encounter with the world's pain—specifically old age, illness, and death—was the immediate stimulus that led to the enterprise of Buddhism, a fact that for our purposes is notable in several respects. For one thing, those phenomena that so troubled the young Buddha-to-be are things that continue to bedevil thoughtful, sensitive people today. And for another, all three—old age, illness, and death—are profoundly and unavoidably natural, and thus deeply biological as well.

The Buddha certainly wasn't unique in being beset with anguish owing to the unavoidable materiality of our lives. In Shakespeare's *As You Like It*, "melancholy Jacques" dilates upon aspects of life that are certifiably as we do *not* like it:

"Thus we may see," quoth he, "how the world wags...
And so, from hour to hour, we ripe and ripe,
And then from hour to hour, we rot and rot:
And thereby hangs a tale."

It is a tale that Buddhists know well, and that no one—Buddhist or not—should ignore. For a different response, although one no less melancholy, here is the Irish poet, W. B. Yeats, whose *Sailing to Byzantium* (1928), expressed contrasting sentiment, namely hope of extricating himself from the flesh's decomposition by retreating into art, notably beauty and the allure of artificial (if misleading) permanence:

O sages standing in god's holy fire,
...consume my heart away; sick with desire
And fastened to a dying animal.
It knows not what it is; and gather me
Into the artifice of eternity.
Once out of nature I shall never take
My bodily form from any natural thing,
But such a form as Grecian goldsmiths make
Of hammered gold and gold enamelling.

Shakespeare's Jacques satisfied himself with a series of melancholy observations. When Yeats bemoaned his biological state ("sick with desire / And fastened to a dying animal"), his response was to imagine himself reincarnated as something artificial, albeit beautiful, and presumably eternal. Others react differently. In his now-classic manifesto *Essays from Round River*, the pioneering ecologist and founder of wildlife management Aldo Leopold wrote that to have an ecological conscience is to "live alone in a world of wounds." The Buddha himself did not live alone, although much of his search for enlightenment did in fact involve treading a fundamentally solitary path. Moreover, the *dukkha* that so troubled the Buddha derives from experiences that are common to all sensitive persons, especially those attuned to the world's immense burden of pain.[v]

A story is told about a young mother, devastated by the death of her child, who came to the Buddha seeking relief from her pain. He said he could cure her distress with a magic potion, which required as a special ingredient just a single mustard seed from the home of a family that had never known death. She dutifully went from door to door, and of course, couldn't find any such people. That realization, itself, didn't eliminate her *dukkha*, but by understanding its universality, her own was easier to bear.[vi]

The world Darwin described, examined, and helped explain is the same world that produces those sources of *dukkha* that so troubled the grieving mother and the young Siddhartha Gautama, and that besets all of us in proportion as we acknowledge our unavoidable and shared participation in *pratitya-samutpada*. It is a world in which disease, old age, and death take place, not only for individuals but increasingly for natural communities and whole ecosystems.

At the conclusion of Shakespeare's *Othello*, when that tragic character is taking responsibility for his misdeeds, he urges his listeners, "Speak of me as I am. Nothing extenuate." By the same token, biologists realize that many aspects of the world are less than pleasing, but they exist nonetheless. Accordingly, we are well advised to speak of the world as it is; nothing extenuate. Although nature is often beautiful (sometimes breathtakingly so), the reality is that it is also harsh, unfeeling, arbitrary, unfair, and unethical (or, rather, nonethical). All living things eventually die, many of them horribly. Disease is everywhere, and although old age is comparatively rare in nature, this is simply because most living things die "prematurely"—that is, something else, often something quite gruesome, knocks them off first, thereby sparing them the ravages of old age.

Darwin knew this. He understood that natural selection isn't always "nature red in tooth and claw,"[vii] but that it is nonetheless deeply enmeshed in the often ugly

---

[v] Note that this is not to say that Buddhism is solely immersed in pain, distress, and disappointment; to the contrary, it is profoundly life-affirming. But this affirmation itself occurs in a context of deep awareness and compassion, which in turn cannot be divorced from the literal derivation of *compassion* (literally, "suffering with").

[vi] It is interesting, by the way, to compare this Buddhist tale about confronting death with the traditional Christian one, in which Christ brings the dead Lazarus back to life. In my opinion, the Buddhist version is not only more believable, it also provides a psychologically meaningful way for all people to deal with this universal biological reality without expecting or hoping for a literal miracle.

[vii] A phrase that originated not with Darwin but with the poet Alfred, Lord Tennyson.

struggle for existence. "All organic beings are exposed to severe competition" wrote Darwin, in *The Origin of Species*:

> We behold the face of nature bright with gladness, we often see superabundance of food; we do not see, or we forget, that the birds which are idly singing round us mostly live on insects or seeds, and are thus constantly destroying life; or we forget how largely these songsters, or their eggs, or their nestlings are destroyed by birds and beasts of prey; we do not always bear in mind, that though food may be now superabundant, it is not so at all seasons of each recurring year.

It is worth quoting further from Darwin, whose writing on this point is actually quite clear, although rarely read today:

> I should premise that I use the term Struggle for Existence in a large and metaphorical sense, including dependence of one being on another, and including (which is more important) not only the life of the individual, but success in leaving progeny. Two canine animals in a time of dearth, may be truly said to struggle with each other which shall get food and live. But a plant on the edge of a desert is said to struggle for life against the drought, though more properly it should be said to be dependent on the moisture. A plant which annually produces a thousand seeds, of which on an average only one comes to maturity, may be more truly said to struggle with the plants of the same and other kinds which already clothe the ground.

For Darwin, and for his intellectual descendants among biologists, "there is no exception to the rule that every organic being naturally increases at so high a rate, that if not destroyed, the earth would soon be covered by the progeny of a single pair."

This, of course, doesn't happen. And why not? Because something intervenes between the huge reproductive potential of every organism and the biological reality whereby most living things give rise to far fewer genetic representatives than they are theoretically capable of producing. That "something" is natural selection: differential reproduction among organisms and genes, which results in some organisms leaving substantially more descendants than others. A key point for our purposes is that this process of differential reproduction is likely to be none too pretty. As Darwin wrote in a letter to the botanist Joseph Hooker:[viii] "What a book a Devil's Chaplain might write on the clumsy, wasteful, blundering low & horribly cruel works of nature."

[viii] In a response, as it happened, to Thomas Huxley's inquiry about the possible mating behavior of hermaphroditic jellyfish.

Here and there Darwin tried to soften the hard edges of natural selection: "When we reflect upon this struggle," he opined in *The Origin of Species*, "we may console ourselves with the full belief that the war of nature is not incessant, that no fear is felt, that death is generally prompt." In this regard, we now know that Darwin was mostly wrong. The natural death of animals is often slow and agonizing, filled with *dukkha*.

Darwin took a different and more realistic tack when writing to Asa Gray, a noted American botanist and supporter of evolution, although also a devout Christian:

> I cannot persuade myself that a beneficent and omnipotent God would have designed and created the Ichneumonidae with the express intent of their feeding within the living bodies of Caterpillars,[ix] or that a cat should play with mice.

Our current understanding of natural selection has not softened this conclusion. Living things are selected to maximize their fitness, which is their success in projecting copies of themselves—via their genes—into the future, relative to the success of their competitors. As the noted evolutionary biologist George C. Williams points out, our current understanding of natural selection is that it operates as a ratio, with the numerator reflecting a measure of genetic success and the denominator, the success of alternative genes. Since a gene (or an individual, a population, even—in theory—a species) maximizes its success by producing the largest such ratio, it can do so either by reducing the denominator or increasing the numerator. Most creatures, most of the time, find it easier to do the latter than the former, which is why living things generally are more concerned with feathering their nests than defeathering those of others.

Taken by itself, such self-regard isn't the stuff to cheer an ethicist or to reassure a Buddhist looking for relief from the *dukkha* that is biology's bequest to us all. And to make matters worse, animal studies in recent years have revealed a vast array of behavior whereby living things have no hesitation in minimizing the denominator, trampling over others in pursuit of their own biological benefit. The living world is replete with grisly cases of predation and parasitism, a universe of ghastly horrors all generated by natural selection and unleavened by the slightest ethical qualms on the part of perpetrators.

In her stunning memoir *Pilgrim at Tinker Creek*, Annie Dillard described her dismay at watching a frog whose innards were liquefied and then sucked dry by a giant water bug:

> Just as I looked at him, he slowly crumpled and began to sag. The spirit vanished from his eyes as if snuffed. His skin emptied and drooped; his very skull

---

ix The Ichneumonidae are an insect family of parasitic wasps that insert their eggs into other insects, and which then develop into larvae which devour their "hosts" from within.

seemed to collapse and settle like a kicked tent. He was shrinking before my eyes like a deflating football. I watched the taut, glistening skin on his shoulders ruck, and rumple, and fall. Soon, part of his skin, formless as a pricked balloon, lay in floating folds like bright scum on top of the water: it was a monstrous and terrifying thing. I gaped bewildered, appalled. An oval shadow hung in the water behind the drained frog; then the shadow glided away.

This "shadow," which had just killed the frog—so smoothly, mercilessly, and *naturally*, before gliding away—was an enormous, heavy-bodied brown creature that

eats insects, tadpoles, fish, and frogs. Its grasping forelegs are mighty and hooked inward. It seizes a victim with these legs, hugs it tight and paralyzes it with enzymes injected during a vicious bite.... Through the puncture shoot the poisons that dissolve the victim's body, reduced to a juice. This event is quite common in warm fresh water.... I stood up and brushed the knees of my pants. I couldn't catch my breath.

Dillard quickly goes on, achieving a more objective and scientific detachment:

Of course, many carnivorous animals devour their prey alive. The usual method seems to be to subdue the victim by downing or grasping it so it can't flee, then eating it whole or in a series of bloody bites. Frogs eat everything whole, stuffing prey into their mouths with their thumbs. People have seen frogs with their wide jaws so full of live dragonflies they couldn't close them. Ants don't even have to catch their prey: in the spring they swarm over newly hatched, featherless birds in the nest and eat them tiny bite by bite.

She notes, with understatement, that "its rough out there, and chancy," that "every live thing is a survivor on a kind of extended emergency bivouac," and that "cruelty is a mystery," along with "the waste of pain." And finally, Ms. Dillard concludes that we "must somehow take a wider view, look at the whole landscape, really see it, and describe what's going on here. Then we can at least wail the right question into the swaddling band of darkness, or, if it comes to that, choir the proper praise."

Dillard also observes that "parasitism...is a sort of rent, paid by all creatures who live in the real world," and she concludes, ruefully, that in acknowledgment of such reality, "Teddy bears should come with tiny stuffed bear-lice." If, in Christian tradition, the wages of sin are death, in the world of biological beings, the wages of life are parasitism, predation, and plain old-fashioned murder.

In Stephen Sondheim's dark musical *Sweeney Todd*, we learn similarly that "the story of the world, my sweet, is who gets eaten and who gets to eat." In the nonfiction world we all inhabit, there is pleasure and pain, suffering and delight, eaten and eater, life and death, growth and decay. Nonetheless, it is surprisingly easy, especially in modern urban areas, for people to go for extended periods without ever encountering death. Even in natural environments, diseased or dead animals are comparatively rare, mostly because the diseased don't survive long and the deceased are quickly consumed and/or decomposed. But this is not to say that disease, old age, and death have in any sense been vanquished!

In fact, by extending the average human lifespan without eliminating disease, old age, and death, modern technology has created a more expanded playing field, providing new opportunities for *dukkha*. There has certainly been progress in public health, for example, and yet this has largely involved substituting new diseases for old (fewer lethal epidemics, such as smallpox or typhoid; more debilitating illnesses—often complexly linked to environmental pollution—such as cancer, heart disease, diabetes, etc.), as well as increasingly extended periods of old age and therefore senility, dementia, and other illnesses of the elderly. These chronic conditions have become the new epidemics, but whereas smallpox, typhus, typhoid, cholera, and other infectious diseases killed relatively quickly, these scourges don't.

Worse yet, perhaps, are those entirely "natural" cases of vicious genetic self-promotion at the expense of others. For example, biologists have documented infanticide in numerous species, including lions and many nonhuman primates such as langur monkeys and chimpanzees. The basic pattern is that when the dominant male in a harem group is overthrown, his replacement often systematically kills the nursing infants, who are unrelated to the victor, thereby inducing the lactating mothers to resume their sexual cycling, whereupon they mate with their infant's murderer. It is truly awful, such that even hard-eyed biologists had a difficult time accepting its ubiquity and even, until recently, its "naturalness." But natural it is, and also a readily understood consequence of evolution as a mindless, automatic and value-free process, whose driving principle is if anything not just amoral but—by any decent human standard—downright immoral.

Add cases of animal rape, deception, nepotism, siblicide, matricide, and cannibalism, and it should be clear that natural selection has blindly, mechanically, yet effectively favored self-betterment and self-promotion, unmitigated by any ethical considerations. I say this fully aware of an important recent trend in animal behavior research: the demonstration that animals often reconcile, make peace, and cooperate. And yet, no less than the morally repulsive examples just cited, these behaviors also reflect the profound self-centeredness of the evolutionary process. If the outcome in certain cases is less reprehensible than outright slaughter, it is only because natural selection only sometimes works to reduce the denominator of the fitness ratio. Most of the time, it increases the numerator.

But all the time, the only outcome assessed by natural selection is whether a given tactic works—whether it enhances fitness—not whether it is good, right, just, beautiful, admirable, or in any sense moral. Why, then, should we look to such a process for moral guidance? Indeed, insofar as evolution has engendered behavioral tendencies within ourselves that are callously indifferent to anything but self- (and gene-) betterment, and armed as we now are with insight into the origin of such tendencies, wouldn't sound moral guidance suggest that we intentionally—mindfully, as modern Buddhists like to put it—act contrary to them?

In the movie *The African Queen*, Katherine Hepburn stiffly observes to Humphrey Bogart: "Nature, Mr. Allnut, is what we were put on earth to rise above." I strongly doubt that we were put on earth to do anything in particular (see chapter 7), but if we want to be ethical—rather than simply successful, at least biologically speaking—rising above our human nature may be just what is needed. Evolution by natural selection, in short, is a wonderful thing to learn *about*—but a terrible thing to learn *from*.

By the end of the nineteenth century, Thomas Huxley was perhaps the most famous living biologist, renowned in the English-speaking world as "Darwin's bulldog" for his fierce and determined defense of evolution by natural selection. But he defended it as a scientific explanation, not a moral touchstone. In 1893 Huxley made this especially clear in a lecture titled "Evolution and Ethics," delivered to a packed house at Oxford University:

> The practice of that which is ethically best—what we call goodness or virtue—involves a course of conduct which, in all respects, is opposed to that which leads to success in the cosmic struggle for existence. In place of ruthless self-assertion it demands self-restraint; in place of thrusting aside, or treading down, all competitors, it requires that the individual shall not merely respect, but shall help his fellows; its influence is directed, not so much to the survival of the fittest, as to the fitting of as many as possible to survive.

"The ethical progress of society depends," according to Huxley, "not on imitating the cosmic process [that is, evolution by natural selection], still less in running away from it, but in combating it."

It may seem impossible for human beings to "combat" evolution, since *Homo sapiens*—no less than every other species—is one of its products. And yet, Huxley's exhortation is not unrealistic. It seems likely, for example, that to some extent each of us undergoes a trajectory of decreasing selfishness and increasing altruism as we grow up, beginning with the infantile conviction that the world exists solely for our personal gratification and then, over time, experiencing the mellowing of increased wisdom and perspective as we become aware of the other lives around us, which are not all oriented toward ourselves. In her novel *Middlemarch*, George Eliot noted that "we are all born in moral stupidity, taking the world as an udder with which to feed ourselves." Over time, this "moral stupidity" is replaced—in varying degrees—with ethical acuity, the sharpness of which can largely be judged by the amount of unselfish altruism that is generated.

One of the most important Buddhist teachings, the Metta Sutra, develops a very similar lesson. Recited daily by Theravadan monks, it includes this admonition:

> Even as a mother protects with her life her child, her only child, so with a boundless heart should one cherish all living beings. Let thoughts of loving kindness for all the world radiate boundlessly, into the sky and into the earth, all around, unobstructed, free from any hatred or ill-will. Standing or walking, sitting or lying down, as long as one is awake, one should develop this mindfulness.

Metta, in Pali (*maitri* in Sanskrit), is often translated as "tender regard" or "loving-kindness," two rather clunky phrases that nonetheless share a delightfully old-fashioned (but never out-of-fashion) flavor.

Many sober, highly intelligent scientists and humanists misunderstand the connection between evolution and morality, grimly determined that evolutionary facts are dangerous because they justify human misbehavior. The developmental psychologist Jerome Kagan exemplifies this blind spot. "Evolutionary arguments," he writes, "are used to cleanse greed, promiscuity, and the abuse of stepchildren of moral taint."[3] Similar arguments were in fact used in this way, in the unlamented days of late-nineteenth-century social Darwinism. But no longer.

Today, they are used to understand greed, promiscuity, bigotry, parent-offspring conflict, and the abuse of stepchildren—while they also help us understand parenting, nepotism, reciprocity, friendship, courtship, and *metta*-like altruism, to name just a few. Like the famous yin-yang symbol, biology contains nasty and aversive elements inextricably entwined with beautiful and inspiring ones. Leave it to the Buddhists to recognize the interconnectedness of both beauty and nastiness in everyday life. For instance, the much-loved lotus flower (*Nelumba nucifera*) is a popular Buddhist symbol. An aquatic perennial, the lotus thrives in stagnant swamps and marshes, where, out of the stinking mud and muck, it generates a large and gorgeous flower.

For all the similarities, parallels, and convergences that characterize Buddhism and biology (and I trust that by now, the reader has been pretty much convinced on this point), there are also some notable differences, including the easily overlooked fact that biology—like other sciences—makes use of objectively quantified data, whereas Buddhism employs the subjective input of personal experience. In addition, although biology and Buddhism both acknowledge the existence and importance of *dukkha* in the natural world, it is important to note that biology, as a science, does not contain ethical prescriptions, whereas Buddhism, as a _____ (fill in the blank as you choose: religion, philosophy, practice tradition), definitely does. Or at least, Buddhism as practiced by many of its adherents in the twenty-first century definitely does.

As we have emphasized, however, early Buddhism was not especially "engaged"— beyond the Buddha's effort to engage disciples in their own enlightenment, even though this enlightenment occurred when he perceived the universality of human suffering. Compared especially with the many people who even today look the other way when confronted with such sources of *dukkha*, the Buddha's concern was in fact extraordinarily engaged.

Nonetheless, early doctrine claims that the Buddha's actual awakening—as distinct from his perceiving of *dukkha*—occurred when he became aware of the "dependent arising" (our old friend, *pratitya-samutpada*) of all of existence, which was initially seen, therefore, as a problem rather than something to celebrate. In one of the primary texts of Buddhism, the Three Characteristics of *Samsara*—the supposed unending worldly flow of birth, death, and rebirth—consist of failure to recognize that all the world's constituents are impermanent or transitory (*anitya*), lacking in an independently existing inner self (*anatman*), and therefore ultimately unsatisfying (*dukkha*). In his famous First Sermon, the Buddha states:

> Now this, monks, is the Noble Truth of *dukkha*: birth is *dukkha*, aging is *dukkha*, death is *dukkha*; sorrow, lamentation, pain, grief, and despair are *dukkha*; association with the unbeloved is *dukkha*; separation from the loved is *dukkha*; not getting what is wanted is *dukkha*. In short, the...aggregates to which one clings are all *dukkha*.

In *The Light of Asia* (1879), the British poet Edwin Arnold (1832–1904) depicted the Buddha as deeply saddened by the dog-eat-dog realities of *dukkha* engendered by biological competition:

> ...looking deeply, he saw...
> How lizard fed on ant, and snake on him,

And kite on both; and how the fish-hawk robbed
The fish-tiger of that which it had seized;
The shrike chasing the bulbul, which did chase
The jeweled butterflies: till everywhere
Each slew a slayer and in turn was slain,
Life living upon death....
The Prince Siddhartha sighed.

The solution, according to early Buddhism, is to see through the various passionate clingings and cravings in which we and all other living things are entangled and that produce so much suffering, to transcend avid welcoming of and yearning for the things of this world and to avoid fervent hatreds thereof, to renounce sensual pleasures, and to achieve liberation from and nonattachment to the troublesome "realities" of existence. This is almost the precise opposite of engagement as represented in modern times by the work of the Dalai Lama, Thich Nhat Hanh, and a number of Western masters such as Joanna Macy and Robert Aitken.

Buddhist engagement nonetheless has a lengthy and noble history as well. The eighth-century Indian scholar and philosopher Santideva provides us with perhaps the earliest clear statement of deriving ethical precepts from the Buddhist teaching of *pratitya-samutpada* as we have explored it: "Just as the body, which has many parts owing to its division into arms and so forth, should be protected as a whole, so should this entire world, which is differentiated and yet has the nature of the same suffering and happiness."[4] David McMahan points out that similar statements can be found in East Asian texts, such as the writings of the early-fifteenth-century Korean monk Gihwa:

> Humaneness implies the interpenetration of heaven and earth and the myriad things into a single body, wherein there is no gap whatsoever. If you deeply embody this principle, then there cannot be a justification for inflicting harm on even the most insignificant of creatures.[5]

It can certainly be argued that the Buddha's original teaching was concerned not so much with helping us feel more connected, but less. For our purposes, however, this misses the point. We—I and, I assume, most readers of this book—are less interested in unearthing the "true" and "original" teaching of the Buddha (or of anyone else) than in gaining wisdom wherever it can be found. And there is no reason to think that more such wisdom is present in truly "authentic" teachings when such authenticity is simply a function of antiquity.

The Buddha's original concerns are, if anything, highlighted and italicized in the modern world, which has added its own unique sources of *dukkha*. In addition to the chronic diseases of old age and material abundance described earlier, these new problems include the possibility of nuclear mega-death, as well as the current realities of widespread pollution, global climate change, rampant consumerism, the degradation of biodiversity, and so forth. The *dukkha* that upset the Buddha 2,500 years ago is heightened—certainly not reduced—in the twenty-first century. For the Buddha, the downsides of human life are not a punishment for our misdeeds, not something visited upon us by an angry deity; rather, they are simply the result of how things are. For modern, engaged Buddhists, they are also the result of what we have done and what we are now doing.

Interestingly, there do not appear to be any words in Pali or Sanskrit (the two foundational languages for early Buddhist writings) that employ the word "nature" in anything approaching its modern English usage. So we need to be careful in describing Buddhism as inherently biofriendly. Nonetheless, appreciation of the natural world is prominent in Buddhist teaching, especially as it was influenced by Taoist perspectives after Buddhism arrived in China.

Dharma—an important part of Buddhist thought—is the way things are, the underlying nature of things, as well as the teaching of the Buddha regarding these things. Yet *dharma* is not only the way the world is,ˣ but the way we ought to behave if we wish to minimize harm to ourselves and others. As a key element of *dharma*, Buddhist compassion is thus not a willed activity, something forced against one's inclination. Rather, it happens in the same way that your heart beats or your lungs breathe. It is altogether natural, if we only have enough insight into our connection to what is mistakenly called "others" ... not only our friends and relatives but everyone, including those who might seem to be our opponents. "If we could read the secret history of our enemies," wrote Henry Wadsworth Longfellow, "we should find in each life sorrow and suffering enough to disarm any hostility."[6]

As we have seen, the first Buddhist precept is *ahimsa*, or non-harming, a teaching that has long been interpreted to refer to living things generally, and not just to other people. The English word "compassion," so crucial to Buddhist practice, is significant in its derivation as well as its meaning. It originates in the word "passion," which refers not to erotic desire but "suffering," as in Christian references to the "passion of Christ." Com-passion therefore bespeaks "suffering with" and is intimately related to an awareness of being connected—awareness, that is, of *anatman, anitya,* and *pratitya-samutpada.*

"Suffering with," however, has not always translated into "identifying with," especially among people reluctant to acknowledge their connectedness. This is especially pronounced when it comes to the matter of human beings as animals. "In an

---

ˣ Thus, not entirely distinct from the *tao* of Lao-tse.

aversion to animals," wrote the early-twentieth-century philosopher and cultural critic Walter Benjamin, "the predominant feeling is fear of being recognized by them through contact. The horror that stirs deep in man is an obscure awareness that in him something lives so akin to the animal that it might be recognized." For Benjamin, at least, the widespread fear of being connected to animals is due in large part to an often unacknowledged awareness that there actually is a sound basis for making such a connection, an awareness found even in the most fervent, evolution-denying religious fundamentalists.

Although Buddhism is unquestionably more bio-friendly than the Abrahamic big three, it really isn't nature worship as such, especially not in its earlier, formative stage. Traditional Buddhist teaching identifies a hierarchy in nature, with animals definitely lower than humans, although connected (and in this sense, Buddhism is far more inclined to situate people in nature than is Judaism, Christianity, or Islam). At one point, for example, the renowned fifth-century Buddhist commentator and sage Buddhaghosa notes that there is more "demerit" in killing a large animal than a small one, so that killing an elephant, for example, generates a greater karmic burden than does killing a mouse, which in turn is more burdensome than killing a fly. But this isn't so much because the elephant is especially valued relative to the mouse, or the mouse relative to a fly, but rather because it takes more energy and thus intention to kill a larger animal. In addition, monastic Buddhist law holds that killing a human being is a far greater offense than is killing an animal—even though Buddhists also proclaim that any creature could have been one's great-great-grandmother.

One of the most direct routes to genocide and war has long been dehumanization of the enemy via animal images: it is easier to kill those who have first been reduced to rats, dogs, bugs, or other "vermin." To be sure, people sometimes derive positive self-images from animals: no human society is without admiring references such as soaring like an eagle, being brave like a lion, strong as a bull, busy as a bee, and so forth. But it is one thing to be *like* an animal; it is quite another to face the fact of *being* an animal, we and they deeply and irrevocably united via our anatomy, physiology, behavioral inclinations, even our literal biological ancestry.

In "The Individual, Society and Nature," Thich Nhat Hanh goes beyond the teaching of earlier Buddhist masters and urges us to face those facts of intimate connectedness, reminding us that having done so, we are impelled to behave differently as a consequence:

A human being is an animal, a part of nature. But we single ourselves out from the rest of nature. We classify other animals and living beings as nature, acting as if we ourselves are not part of it. Then we pose the question, "How should we deal with Nature?" We should deal with nature the way we should deal with ourselves!

We should not harm ourselves; we should not harm nature. Harming nature is harming ourselves, and vice versa....Human beings and nature are inseparable. Therefore, by not caring properly for any one of these, we harm them all.[7]

Here we see an important point: modern, engaged Buddhism involves not only recognizing our connectedness to the natural world, but also encouraging a responsibility to care properly for our fellow creatures as well as their and our environment. It is modern and thus a new wrinkle in Buddhism, but one that has discernible antecedents.

"I came to realize clearly," wrote Dogen, the thirteenth-century founder of the Japanese Soto school of Zen Buddhism, "that mind is not other than mountains and rivers and the great wide earth, the sun and the moon and the stars."[8] This stunningly deep sense of responsibility comes down to—and derives from—acknowledging the reality of *anatman*, the lack of distinction between a separate self inside and the rest of the world outside, along with its companion concepts, *anitya* and *pratitya-samutpada*. At the same time, it needs to be emphasized that unlike some traditions in the West, especially recent ones, Buddhism doesn't romanticize nature, partly because a Rousseauian "return to nature" makes no sense if we never left. This is especially true of the Theravadan branch. By contrast, the Chan (Chinese) offshoot is more overtly nature-friendly, probably because of the preexisting Taoist admiration for nature. This, in turn, influenced the Zen (Japanese) tradition, including Dogen as well as others.

Buddhism is notably immersed in the natural world, perhaps more than is any other religious/spiritual/practice tradition (with the possible exception of Taoism). This nature-friendliness is especially true of Buddhism's East Asian manifestations, namely Chan/Zen. Before being exposed to Buddhism, East Asian traditions did not entertain the notion of a reality beyond that of immediate, natural existence; as a result, when Buddhism was incorporated into these early Taoist and naturist traditions, the result was distinctly more nature-friendly than was characteristic of Buddhism informed by the Pali Canon.

Nonetheless, Indian Buddhism provided fertile ground for accepting, emulating, and admiring nature, a perspective that is only a short step away from seeking to preserve and defend it. It is no coincidence that the Buddha supposedly gained enlightenment while sitting under a tree. Mahakashyapa, a major disciple in the Buddha's own time, extolled the joy of experiencing nature, saying that it was exceeded only by the pleasure of true insight. According to legend, in his Flower Sermon the Buddha held up a white flower and silently admired it, whereupon his audience was perplexed—except for Mahakashyapa, who smiled. The Buddha concluded from this that Mahakashyapa best understood his message and chose him as the first disciple, worthy to be the Buddha's successor.

Subsequently, numerous Buddhist seekers have looked to nature for sustenance, inspiration, and wisdom. The Chinese poet Han-shan (believed to have flourished in the eighth or ninth century) wrote:

As for me, I delight in the everyday Way,
Among mist-wrapped vines and rocky caves.
Here in the wilderness I am completely free,
With my friends, the white clouds, idling forever.

The implication—quite specific in many Buddhist texts—is that nature is correct, the way things ought to be.[xi]

Speaking here and now in modern times, Thich Nhat Hanh recommends that each of us meditate on our overall connectedness to nature as well as to other people, urging his listeners to work at this "until you see yourself in the most cruel and inhumane political leader, in the most devastatingly tortured prisoner, in the wealthiest man, and in the child starving, all skin and bones." He goes on, encouraging everyone, whether Buddhist or non-Buddhist, to

practice until you recognize your presence in everyone else on the bus, in the subway, in the concentration camp, working in the fields, in a leaf, in a caterpillar, in a dew drop, in a ray of sunshine. Meditate until you see yourself in a speck of dust and in the most distant galaxy.

Since we now realize that "one is all, all is one" in our bodies, let us go another step and meditate on the presence of the entire universe in ourselves. We know that if our heart stops beating, the flow of our life will stop, and so we cherish our heart very much. Yet we do not often take the time to notice that there are other things, outside of our bodies, that are also essential for our survival. Look at the immense light we call the sun. If it stops shining, the flow of our life will also stop, and so the sun is our second heart, our heart outside of our body. This immense "heart" gives all life on earth the warmth necessary for existence. Plants live thanks to the sun. Their leaves absorb the sun's energy, along with carbon dioxide from the air, to produce food for the tree, the flower, the plankton. And thanks to the plants, we and other animals can live. All of us—people, animals, and plants—"consume" the sun, directly and indirectly. We cannot begin to describe all the effects of the sun, that great heart outside of our body.

In fact, our body is not limited to what lies inside the boundary of our skin. Our body is much greater, much more immense. If the layer of air around our

---

xi Which leads to an interesting question: is *dukkha* included among these "oughts"?

earth disappears even for an instant, "our" life will end. There is no phenomenon in the universe that does not intimately concern us, from a pebble resting at the bottom of the ocean, to the movement of a galaxy millions of light years away. The poet Walt Whitman said, "I believe a leaf of grass is no less than the journeywork of the stars...." These words are not philosophy. They come from the depths of his soul. He said, "I am large, I contain multitudes."[9]

Being large and containing multitudes is a familiar sensation among Buddhists, and so, increasingly, is the sense of responsibility that results. Sometimes the outcome might appear distinctly un-Buddhistic in its feistiness. An old and beloved Buddhist prayer, the Flower Ornament Sutra, goes:

When I see flowing water
I vow with all beings
To develop a wholesome will
And wash away the stains of delusion.[10]

Aware of this precedent, the American Zen master and environmental activist Robert Aitken put together his own modernized vow:

With tropical forests in danger
I vow with all beings
to raise hell with the people responsible
and slash my consumption of trees.[11]

From a "wholesome will" to "raising hell" might seem like a big leap, but not necessarily, especially for those who acknowledge the necessity of active compassion, whether from a Buddhist perspective or a biological one.

There is a story—perhaps apocryphal—in which the American avant-garde composer John Cage was asked by his friend, the abstract expressionist artist Robert Rauschenberg, whether he thought that there was too much suffering in the world. "No," Cage supposedly replied, "I think there's just the right amount." I must confess to a substantial helping (perhaps more than the Buddhistically appropriate "right amount") of anger at this statement on the part of the Zen-influenced Cage. His point should be clear: Mr. Cage was aligning himself with the Buddhist concept of balance. My fear, however, is that too much balance of this sort leads to a kind of suffocating equanimity that risks degenerating into self-satisfied inaction in the face of pain and suffering.

I far prefer the story told by the anthropologist and essayist Loren Eiseley,[12] subsequently rewritten in various ways. Someone is walking along a beach that is littered with exposed starfish who are dying in the low tide. He sees a young woman who carefully picks one up and flings it into the ocean. "Young lady," says the observer, "don't you realize there are miles and miles of beach, and starfish everywhere? You can't possibly make a difference." She listened politely, then picked up another and tossed it into the water, saying, "It makes a difference to *this* one."

Buddhist tradition includes among its earliest written literature numerous *jatakas*, dating from about the fourth century BCE. These are legendary stories about the childhood and earlier incarnations of the Buddha; many—perhaps most—of them involve him taking the form of an animal and then often sacrificing ("nobly giving") himself for the sake of others, often other animals. If not quite raising hell on behalf of other life forms, the *jatakas* certainly depict a committed sense of oneness "with all beings." In different *jatakas*, the Buddha is variously described as a "deer king" who frees all creatures from mortal danger, a parrot who selflessly flies through a forest fire to rescue those trapped inside, a monkey who saves his troop from a prince who was also a thoughtless hunter, and a rabbit who gives of his own flesh to a famished tigress so that she in turn can nurse her starving cubs. Self-sacrifice might be seen as a uniquely Buddhist way of raising hell, although Buddhist modernists such as Aitken are increasingly comfortable with direct political activism.

For our purposes, one take-home message of the *jatakas* is not only that empathy and loving-kindness are desirable traits, worthy of being emulated, but also that the Buddha (and, by extension, the rest of us) is so closely connected to the rest of the living world that to sacrifice on its behalf is not only noble but, in a sense, no sacrifice at all. The Buddha's essence was no different from the essence of other living things and, in the event, no different from that of the rest of us—except that he realized it and most of us do not.

Once again, it does not detract from the biological sophistication and engagement of contemporary Buddhists to note that although there are distinct bio-friendly precursors in the older Buddhist traditions, modern engaged Buddhism is to some extent a recent phenomenon. Most of the early Buddhist *sutras* and other texts, even as they commemorate natural environments, contain little or no praise of wilderness for its own sake, but rather as a place or situation within which human beings can maximize their greatest potential. Thus, the many examples in which nature is admired—sometimes extravagantly—are usually developed for their instrumental, human value more than for the sake of nature as such. The Buddha left a palace to live in the forest, he obtained enlightenment under a tree, and some

Buddhist traditions make much of the value of renouncing civilization to live as a hermit, immersed in nature, though this is almost exclusively for the hermit's sake, not nature's. Indeed, as in Christianity, there is a strand in Buddhism that critiques and devalues the world, seeing it as something to escape from and thus less deserving than many "deep ecologists" would prefer. Nonetheless, this standoffishness has been changing rapidly of late, a transition that is not only much needed but that has been readily achieved not only because suitable antecedents exist, but also because a conceptual tradition that relies heavily on *anatman, anitya,* and *pratitya-samutpada* lends itself seamlessly to precisely such a stance.

Recent statements of Buddhist-inspired ecoconnectedness come, for example, from the longtime pro-environment, antinuclear activist and Buddhist teacher Joanna Macy, who urges that in place of the traditional narrow view of each individual as separate and isolated, we substitute a recognition of "wider constructs of self-identity and self-interest—by what you might call the ecological self or eco-self, co-extensive with the other beings and the life of our planet."[13]

To be sure, you don't have to be Buddhist to agree with these sentiments. It is probably no coincidence, however, that the three best-known modern Buddhists are also renowned as activists and political dissidents: Thich Nhat Hanh (who settled in France, having become persona non grata in his native Vietnam for his passionate opposition to both sides during the Vietnam War), the fourteenth Dalai Lama (in exile in India, having fled Tibet after the Chinese takeover), and Aung San Su Kyi (another Nobel Peace Prize winner, who lived for decades under house arrest in Burma for her embrace of democracy). Engaged Buddhism espouses many different social and political issues, nearly all of them left-leaning: environmental protection, opposition to rampant consumerism, support for pacifism or at least antimilitarism and antiwar stances generally, social justice, and so on. There have been exceptions, notably the enthusiasm on the part of Japanese Zen masters for that country's prosecution of World War II. By and large, however, the convergence of engaged Buddhism with progressive—sometimes radical—politics has been almost as thorough as its convergence with biology. And this, in turn, is likely due to the Buddhist focus on interconnectedness, which corresponds nicely to a similar concern—largely at the societal level—on the part of secular progressives.

Consider, for example, the comment by the severely conservative (and distinctly un-Buddhist) British prime minister Margaret Thatcher that

there is no such thing as society. There are individual men and women, and there are families. And no government can do anything except through people, and people must look to themselves first. It's our duty to look after ourselves and then, also to look after our neighbor. People have got the entitlements too much in mind, without the obligations. There's no such thing as entitlement, unless someone has first met an obligation.[14]

By the same token, there is no way that Ayn Rand, patron saint of right-wing individualism and apostle of selfishness, could ever have been mistaken for a Buddhist. Political liberals, progressives, and socialists, by contrast, embrace the role of a caring, connected, and interdependent society as paramount, and with it a commitment to social responsibility. We have already seen numerous examples of Thich Nhat Hanh's engagement on behalf of social and environmental betterment, while that of the Dalai Lama is legendary. For the American Joan Halifax, who has long been deeply involved in providing hospice and other healing services in developing countries such as India, the Buddhist approach is about "radical intimacy with the world," which results in a "life grounded in kindness, compassion, wisdom and skillful means."[15] It is not simply a semantic game to point out that the Buddhist goal of becoming aware of one's selflessness corresponds to the need for less selfishness in our behavior toward each other and toward our environment.

The U.S. Declaration of Independence would seem an endorsement of separation over collectivity: "life, liberty and the pursuit of happiness" appears to capture individualism in a nutshell. But the Declaration's author was not a greed-is-good guy. "Self-love," Thomas Jefferson wrote to a friend thirty-eight years later,

> is no part of morality. Indeed it is exactly its counterpart. It is the sole antagonist of virtue, leading us constantly by our propensities to self-gratification in violation of our moral duties to others.[16]

Starting especially in the 1950s, the Beats (including Allen Ginsberg, John Cage, Jack Kerouac, and Gary Snyder), were drawn to Buddhism for many reasons, not least its ready association with social and political protest against what they saw as a stifling Western culture of conformity, consumerism, and—most important—a death-dealing capability in the form of environmental destruction, culminating in nuclear weapons.

I do not think the Buddha would conclude that the world contains "just the right amount" of suffering. Nor should we.

Much depends on our ability to see the world as an interconnected whole, from organisms and ecosystems to social networks and human systems. The import of a Buddhistic or Buddhist-inspired worldview—one based on a vision of *anatman*, *anitya*, and *pratitya-samutpada*—is huge, encompassing biology but going even further. And this is true whether one subscribes to traditional Buddhism ("version 1.0") or its modernized, fully engaged, and updated "version 2.0."

Legend has it that when the Buddha gained enlightenment, sitting under that bodhi tree, he was challenged by the evil god Mara, who demanded to know by what authority he was proceeding. To this, the Buddha responded by touching the ground.[xii] And, we are told, the Earth roared its approval.

[xii] This image, with the Buddha depicted touching the ground or, alternatively, pointing to it commonly appears in Buddhist iconography. Interestingly, it parallels Raphael's famous painting *The School of Athens*, in which Plato points upward (toward his concept of ethereal "forms"), while Aristotle points downward, toward reality.

6

Engagement, Part 2 (*Karma*)

"ALL ACTIONS CREATE consequences," wrote the Dalai Lama. "Buddhists call this the law of karma or the law of cause and effect."[1] This is a stripped-down, more basic description of karma than we generally use in the West, where karma is taken to mean a kind of personal moral storage bin into which previous actions are tossed and which then emerge to brighten or bedevil one's future. If karma is the accumulated load of prior behavior, then my karma is necessarily different from yours (something of a problem, admittedly, for a perspective that empha-sizes interconnectedness and the absence of a uniquely distinct self). But if we take the Dalai Lama's understanding—and who better to represent a Buddhist perspective?—karma is more generally applicable, a fundamental law of the uni-verse that is neither more nor less than a straightforward and scientifically valid acknowledgment that actions have consequences: that is, the law of "cause and effect."

Note that the Dalai Lama was talking about "the *law* of karma," rather than karma itself. The reality, however, is that whether law or "thing," the formal Buddhist con-ception of karma is not one that scientists in general or biologists in particular are likely to accept. Nor should they.

As we have seen, a kind of bottom-line, bare-bones reincarnation occurs in the literal recycling of atoms and molecules, fundamental to the biological (and Buddhist) acknowledgment that individuals do not have intrinsic exis-tence, separate and distinct from the rest of the world. But this is a far cry from the more traditional understanding of reincarnation, East and West, whereby not just atoms and molecules but some—typically unspecified—aspect of an individual is reborn into a different body, yet mystically still constituting an ineffable, nonmaterial component derived from his or her prior life, or rather, lives: a soul.

It is unthinkable for traditional Buddhists, and indeed for most followers of the Abrahamic big three, to deny the existence of a soul. But it is equally unthinkable, I assert, for any scientist to accept its existence—that is, to believe seriously in something that is immaterial, eternal, immeasurable, undetectable by any known means, and also complexly and indelibly associated with every one of us, distinct from each other. When I die, my carbon, hydrogen, oxygen, nitrogen, phosphorus, and so forth will be recycled into other creatures, other components of this planet and the universe. I accept this eventuality; indeed, I welcome it.[i] But I cannot accept the fairy tale that I, like some sparkly Tinker Bell, will in any meaningful holistic sense be reborn, reincarnated, inserted, or in any way incorporated into a new, temporary body, and, moreover, that the outcome—what kind of body I will next inhabit—is a direct (karmic) consequence of how well or poorly I have lived my current life. Add to this the standard interpretation of karma—not only a guaranteed future comeuppance but also that my current instantiation has itself in some way been generated by the behavior or misbehavior of my antecedents—and the notion becomes even more scientifically and biologically ludicrous.

I have no difficulty, however, describing Tenzin Gyatso (born Lhamo Dondrub) as the fourteenth Dalai Lama, so long as this means that he is the fourteenth person to hold that position, in the same sense that Barack Obama is the forty-fourth president of the United States, with no implication that he is in any way the reincarnation of George Washington.

"Souls cross ages like clouds cross skies," we are told by a futuristic, grammar-challenged shaman in David Mitchell's best-selling, time-bending novel *Cloud Atlas*, "an' tho' a cloud's shape nor hue nor size don't stay the same, it's still a cloud an' so is a soul." I don't believe this for a moment. At the risk of repeating: there is no evidence, none whatever, for the existence of a soul. We all know many immaterial "things" that are nonetheless real: love, beauty, hate, suffering, fear, hope. They exist, but only as our description of attitudes, possibilities, interpretations, and so forth. But the literal existence of a soul—mine, yours, that of the Buddha or Charles Darwin—as somehow physically real yet also metaphysically nonphysical, highly complex and coherent, yet invulnerable and eternal: this is an extraordinary and altogether different assertion, inconsistent with everything that we know from science. As Carl Sagan emphasized, extraordinary claims require extraordinary

---

[i] Just not quite yet! (Echoes of St. Augustine's famous plea: "God, grant me chastity… but not yet.")

evidence, and when it comes to souls, we lack not only extraordinarily powerful evidence, but any evidence whatsoever.

"The" Buddhist attitude toward reincarnation is diverse and—I must add, at the risk of seeming unkind—muddled. Buddhism typically maintains an account of the soul's rebirth that differs from the prevailing Hindu view, which posits a pervasive, worldwide, irreversible, and permanent *atman*. We've seen that, by contrast, the Buddhist perspective involves *anatman*, the explicit absence of any concrete self. Consider as well the fact that, according to Buddhist thinking, *anitya* is also universal, and the notion of a distinct and unchanging self that is transmitted from a dead or dying body into a new one is simply not tenable. The Buddha described an alternative process analogous to a sequence in which successive candles are lit by the flame of a preceding one; as a result, the array of flames is causally linked, forming a continuing stream, but these flames are not identical. Sometimes, therefore, the term "transmigration" is used instead of "reincarnation."

Nonetheless, it is clear that most Buddhists (and even entire branches of Buddhism, such as the Tibetan) rely heavily on something very much like what most Westerners have come to understand by reincarnation. Many Buddhists claim, for example, that especially enlightened practitioners can remember their past lives.[ii] Call it what you will, such doctrines inevitably assume that souls, although immaterial, are nonetheless somehow real, and to my biologic and scientific mind, they are therefore immediately disqualified from serious consideration. When it comes to proving the existence of a soul, *sutra* citing is no more persuasive than is Bible beating.

Let's face it: reincarnation (or transmigration) is a real problem for any effort to reconcile—or even to point out parallels between—Buddhism and biology. There is, nevertheless, a kind of biological, chemical, and physical legitimacy to the Buddhist assertion of reincarnation—if and only if we give up this silliness about souls. We've already noted that at the subatomic, atomic, molecular, and genetic levels, we are composed of the same stuff that previously made up other creatures, and will do so again, and again (repeat indefinitely). This may not be especially meaningful at the lowest levels—molecular, atomic, subatomic—since, as far as we know, there is no difference among, say, individual carbon atoms or protons, such that mine might be distinct from anyone else's and therefore uniquely associated with those of my immediate ancestors, and eventually with my descendants. (Alternatively, it is always

---

[ii] If, as is widely acknowledged, the test of science is whether a given proposition is falsifiable, the allegation of remembering previous incarnations might well take the cake as the world's least falsifiable. (It is also interesting that such claims often involve having been a king, queen, or other such important personage, whereas simple probability would suggest that the average reincarnation recaller was, in previous lives, well, average.)

possible that this lack of specificity is itself part of the deeper point of "reductionist reincarnation" as I am describing it.)

Things are intriguingly different, in any event, at the level of genes, since those that constitute any individual are quite literally the consequences of their ancestors' bio-karma: those genes that did comparatively well are disproportionately represented in their current physical embodiment as a whole organism. Evolution—the fundamental biological process that generates and is defined by physical continuity over time—is thus nothing but a highly specified process of transmitting a kind of gene-based karma. It must be emphasized, however, that the key to biological success is simply the consequence of genes insofar as their action results in more or less representation in future generations: more success results in more representation, which is to say, higher fitness or being selected for. Less success means less representation, with all that this implies: lower fitness and being selected against. In neither case, however, does it matter whether these same genes contributed to conduct that was more or less Buddhistically meritorious, unless one equates merit with fitness enhancing—and that viewpoint is quite a stretch.

In fact, insofar as they induced bodies to sacrifice themselves on behalf of other, unrelated bodies, merit would in many cases have been selected *against*. Maybe that's what makes certain actions particularly meritorious: that they go against the widespread evolutionary tendency for gene-based selfishness and are therefore worthy of special praise. Praise, perhaps, but success? That's another matter.

In any event, the biological consequences of natural selection are doubly powerful, not only because natural selection is the process largely responsible for generating each organism in the first place—via its "karma" (i.e., the consequence of its impact on the lives of each body's ancestors and ancestral genes)—but because it operates directly on the predispositions of each living thing in real time. A key point is that these predispositions can often favor actions that, although selfish at the level of genes, are actually altruistic when viewed as the behavior of bodies. This is because a gene-based perspective perceives that in a sexually reproducing species such as ours, each individual is most closely associated with its own body but is also literally represented to varying degrees in the genes present within other bodies, with the amount of sameness decreasing as we proceed from close genetic relatives to those more distant. Moreover, one of the most potent and exciting insights of modern evolutionary biology, a revolutionary theory that effectively explains much of so-called altruistic behavior, is based on the recognition that when bodies behave benevolently toward certain other bodies, what is really happening is that genes are looking after identical copies of themselves that temporarily occupy the bodies in question.[2]

When evolutionary biologists realized that the biology of altruism could be understood as the result of genes selfishly benefiting themselves in other bodies, they found that the circle of an organism's concern for others mapped directly onto their concern for an extended genetic self. Reading this, Buddhists might take issue with the fact that (r)evolutionary biology conceives a distinction between closer and more distant relatives, whereas their notion of interconnectedness would balk at such differentiation. I would bet, however, that even the most advanced Buddhist sages distinguish their close relatives from strangers and treat the former somewhat more benevolently.

This does not mean, however, that genes alone determine prosocial versus antisocial behavior. If evolution by natural selection is the source of our mind's a priori (instinctive) structures, then in a sense these structures also derive from experience—not just the immediate short-term experience of each developing organism, but also the long-term experience of an evolving population. A biological lineage, after all, is a pattern of flow, and evolution itself is the manifestation of innumerable such patterns, experience accumulated through an almost unimaginable length of time. The a priori human mind—seemingly preprogrammed and at least somewhat independent of personal experience—along with everything else that constitutes each seemingly separate and isolated body, is actually nothing more than the embodiment of experience itself.

We can of course go further, as both biology and Buddhism do, and base a science as well as a worldview on the fact of genetic connectedness—not only among members of the same species, but also among all living things: genes for most fundamental biological processes are very widely shared, and thanks to evolutionary continuity combined with natural selection's favoring of some genes over others, the more fundamental the impact of particular genes, the greater the sharing. All vertebrates, for example, are more than 95 percent karma-connected when it comes to genes that underwrite cellular metabolism, for example. Moreover, the mechanisms whereby genes are integrated into organisms are themselves widely shared, which is why it is possible for biologists to introduce, say, genes for cold resistance found in deep-sea fishes into tomatoes. Our evolutionary bequeathal is, in an almost literal sense, our karma. Since we are mammals, we have a different karma than we would if we were haplodiploid insects such as bees or ants.

The pattern is well known to anyone with a smidgen of biological sophistication: *Homo sapiens* (to take just one species, admittedly not at random) share nearly all their genes with other primates, although more with the other apes than with baboons or macaques. And we share more genes with other mammals than with birds, reptiles, or fish. And more with other vertebrates than with, say, dragonflies or dung beetles. And so on: a pattern of variation in genetic identity, decreasing in intensity as the focus expands, but nonetheless with no qualitative discontinuities. This is karmic continuity indeed: each widening circle, incorporating individuals who are progressively more distantly related, represents a diminishing probability of genetic identity and, accordingly, of evolutionary self-interest.

When it comes to the moral implications of karma, the matter is somewhat more complicated and, if anything, more interesting. Accordingly, there is much to be said for debunking the oversimplified (and widespread) idea that karma is something that gets attached to our selves, a notion that is unreliable on several levels. For one, it presupposes the existence of a separate and independent self. Moreover, it raises the problem of exactly how any kind of karma—good or bad—gets itself attached to something as slippery as the ostensible human soul. Cosmic glue? Magical vibrating tendrils of invisible, suborganismic connectivity? Of course, a similar problem inheres with the Christian and Islamic sense of sin, generally conceived as a kind of semi-indelible stain that is somehow imprinted upon one's soul.

There is also the problem that karma can have a downright ugly side, justifying bad events. Why are some people terribly poor, sick, the victim of accident, crime, or abuse? Well, they must have had terrible karma; in other words, they deserve it because of transgressions in a prior life. Not surprisingly, in some Asian societies, karma has a history comparable to the West's use of social Darwinism to explain (and in the process, justify) the perpetuation of monarchies as well as treading upon the already downtrodden.

In a world of profound inequity, unfairness, and undeserved suffering, I categorically refuse to accept that personal or social justice is somehow woven into the fabric of the world, whereby accumulated "bad karma" reveals itself in the suffering of those who seem innocent but who actually misbehaved in a prior incarnation and are therefore now getting their just deserts. And vice versa, of course, with respect to those born to hereditary wealth and position.

Earlier we visited Thich Nhat Hanh's haunting poem that included, as perhaps its most noteworthy image, the raped girl and the sea pirate. Most startling is how Hanh puts the blame not only on the rapist pirate but also, notably, on the poet himself and, by extension, all of us. By contrast, traditional Buddhist (and Hindu) teaching about karma would place a large share of the responsibility on the young and—by any reasonable standard—innocent victim. I trust that most readers of this book would agree that such a perspective is abhorrent.

But this doesn't mean that Buddhist karma is wholly to be discounted. In fact, I hope to show that the exact opposite is true: karma is real, not as a mystical guiding principle for the reincarnation of souls or some such poppycock, but as something scientifically valid, closer to the Dalai Lama's invocation of the law of cause and effect, and to Thich Nhat Hanh's emphasis on the extent to which our interconnectedness, combined with the relevance of action, demands that we accept responsibility rather than blame the victim. In fact, I believe a strong case can be made that once we step away from its superstitious dimensions, there is no such thing as a karma-free zone, and that in the realm of karma—as with the other foundational Buddhist concepts we have investigated—there is a deep convergence between Buddhism and biology, one that carries profound moral consequences.

As we have seen, evolutionary biologists have gained immense insight into the social interactions of animals (including people) by noting that altruism tracks genetic relatedness—which is to say, it reflects the extent of their underlying sameness. The term "inclusive fitness" has been widely adopted as a description of the primary motivating process whereby natural selection proceeds. The phrase itself is revealing, since it emphasizes the potency of a phenomenon both biological and Buddhist: inclusiveness itself.

Buddhists as well as biologists attach great importance to the interaction of organism and environment, noting that the organism literally makes its environment and vice versa, so that the two are inseparable. Early in the Pali Canon the Buddha declares that our bodies and all of our faculties "should be regarded as [resulting from] former actions [i.e., karma] that have been constructed and intended and are now to be experienced."[3] Evolutionary scientists would have difficulty with the "intended" part, especially when referring to our more distant and small-brained ancestors, not to mention those comparable to modern jellyfish or sea anemones, which essentially lack brains—and thus, presumably, intentionality as well. But otherwise the Buddha's statement stands as a pretty good elaboration of the role of evolutionary history in structuring biological reality as it is currently understood.

This is explained in a famous discussion between the Greek king Menander and the Buddhist sage Nagasena:

"Venerable Nagasena," asked the King, "why are men not all alike, but some short-lived and some long, some sickly and some healthy, some ugly and some handsome, some weak and some strong, some poor and some rich, some base and some noble, some stupid and some clever?"

"Why, your Majesty," replied the Elder, "are not all plants alike, but some astringent, some salty, some pungent, some sour, and some sweet?"

"I suppose, your Reverence, because they come from different seeds."

"And so it is with men! They are not alike because of different karmas...each is ruled by karma, it is karma that divides them into high and low."[4]

The structure, physiology, and behavior of today's living things—all of them, including human beings—result directly from the past actions of their innumerable forebears. Thus, when Buddhist texts refer to the fact that our current existence and inclinations are the result of prior actions accumulated via "countless lifetimes," they are literally correct in the sense of modern biology, since the genotype of every creature currently alive is the practical consequence of the actions of each of its antecedents.

The situation continues and has the potential of becoming even more illuminating—and/or more fraught—when moral choices are at issue. Thus, we are not merely passive recipients of our biological forebears; human beings in particular have the opportunity—for many, the obligation—to create themselves in the process of deciding how to behave (in this regard, there is an intriguing convergence between Buddhism and existentialism, which we consider in the next chapter). *Sankharas*—usually translated as "mental formations"—are, in Buddhist thought, our own personality and character traits, which are in turn formed by our actions. Distinct behaviors lead to distinct results, rendering the two essentially indistinguishable.

The word "karma" in Sanskrit means "action," and in its original usage it commonly referred to actions that supposedly enmeshed the actor in *samsara*, the troublesome wheel of repetitive birth and rebirth. Buddhist practice was initially promoted as a way of overcoming this awkward connection. These days, much Buddhist practice continues to involve karma, but in a rather different way: as an encouragement to mindful involvement in the world.

There are numerous exemplars of this transition. For example, the renowned Indian philosopher, anthropologist, writer, and politician Biro Ramji Ambedkar (1891–1956) was born a Hindu untouchable, later converted to Buddhism, and was influential in refocusing modern Buddhist thought and practice. Ambedkar—much beloved in modern India—actually had little patience for the Four Noble Truths, which he felt placed too much emphasis on one's own ignorance and selfishness as a cause for suffering and didn't give enough attention to the thoughtless, heartless, selfish actions of others and the existence of "structural violence" that kept the poor, poor.

It bears repeating that even though the Buddha's original teachings appear to be more concerned with personal enlightenment than with social responsibility, indications of the latter can be found throughout early Buddhism. For example, the Buddha's final words are translated as variations on "All things decay [*or* are transient]," followed by "Strive with earnestness [*or* diligence]." Even assuming a wide range of potential translations into English, this injunction hardly endorses noninvolvement in the world. Sometimes the more "selfish" phrase "for your own enlightenment" is appended, sometimes not. There can be little doubt, however, that striving with earnestness is a far cry from passive inaction.

At the same time, it would be disingenuous on my part to rely too heavily upon particular examples cherry-picked to make it seem that originalist Buddhism was actually fully concordant with a modern, more socially engaged, "politically correct" Buddhism. For example, although it is possible to make a case for gender equality within Buddhism, it is quite clear that early Buddhism was sexist: the Buddha is said to have agreed to the ordination of nuns only after having been asked many times, and even then he acquiesced with great reluctance, insisting that a nun who had served for a hundred years had to be subordinate to a novitiate monk who had just begun his training that day. Close reading of the early texts makes it clear, moreover, that the only way a woman could obtain nirvana would be after she had been reincarnated as a monk! Yet even though Buddhism as practiced in much of Asia remains distinctly male oriented, the gendered aspect of Buddhism—especially as practiced in the West—has changed dramatically, such that women are as well represented as men. Buddhist modernity, in this respect, is not only altogether appropriate but a long-overdue improvement over its more traditional form.

By contrast, it doesn't require major conceptual reframing to generate engaged Buddhism from Buddhism's deeper truths. Start with the realities of *anatman, anitya,* and *pratitya-samutpada,* mix well, add a dose of karma-as-action, and you get a prescription for direct engagement for the betterment of "all sentient beings."[iii]

To my knowledge, there is no other religion or spiritual tradition that comes close to Buddhism's insistence on taking all "sentient beings" into account. Here are four different versions of the most basic *bodhisattva* vow: "Beings are numberless, I vow to save them" (San Francisco Zen Center); "All beings, without number, I vow to liberate" (Rochester Zen Center); "The many beings are numberless, I vow to save them" (Diamond Sangha, Hawaii); and "Sentient beings are numberless; I vow to save them" (Allen Ginsberg, Gary Snyder, and Philip Whalen).[5] Pledging to do so, incidentally, may appear altogether unrealistic, even preposterously megalomaniacal. And maybe it is. But if you accept that we are all intimately connected and that actions have consequences, the path (*dharma*) is clear.

Engaged Buddhism is thus not so much the annihilation of the self as its extension into additional, otherwise unacknowledged selves. What does it mean to say that you feel the pain of the disappearing rain forests, of an elephant shot for its

---

[iii] Regrettably, such involvement need not always be benevolent. Thus, just as the probability of genetic sameness—the closeness of inherited relationship—predicts altruism in a wide range of species, there is a dark flip side as well: It is at least possible that some of our nasty, exclusionist, and even violent behaviors are a result of the degree to which we are *not* related. After all, the closer the genetic relationship, the more similar the individuals; by the same token, the more distant the relationship, the more different the individuals. This leads to the dispiriting proposition that to some degree, intergroup intolerance and even racism may derive from the ugly alternative to genetic connectedness—namely, perceived separation.

tusks, or a whale butchered for meat? More than anything, it says that you recognize a reality that is as biological, and as karmic, as it is Buddhist.

An intriguing possible neurobiological route to the subjective experiencing of this reality, involving so-called mirror neurons, has recently been discovered.[6] These were first described by Italian researchers in the early 1990s and are located especially in a brain region known as the insula. Mirror neurons evidently involve a mirroring (what else?) within an observing individual of whatever he or she sees happening to someone else. It could even be someone of another species, at least in the case of monkeys. Indeed, the initial, accidental discovery of mirror neurons was made when a rhesus monkey in a lab in Parma, Italy, had an electronic probe inserted into a region of its brain concerned with movement and intended movements. A researcher happened to enter eating an ice cream cone, whereupon the monkey watched him intently; then, as the person raised the cone to his mouth, the monkey's neurons fired in synchrony—or, one might say, in sympathy. Just watching caused the monkey's brain to experience a mirror image of the same movement, identical to that which occurs when the monkey raised a peanut to its own mouth.

Human beings also have mirror neurons, and, not surprisingly, ours are even more precise, flexible, and competent than those of our primate cousins. Mirror neurons are also found in other brain regions, specifically the premotor cortex, the posterior parietal lobe, and the superior temporal sulcus. Although their precise role is as yet unclear, perhaps they act to help us feel, deeply, the actions of our fellow humans. After all, our evolutionary fitness depends on an accurate theory of mind,[iv] and on anticipating the actions of others, as well as being able to imagine what those others are experiencing. They may also facilitate learning from others, an important phenomenon characteristic of most intelligent animals and appropriately known as "observational learning."

Essentially, when we see someone doing a physical action, part of our brain rehearses the same movement. Maybe watching a movie therefore isn't entirely vicarious, and maybe virtual reality is less virtual and more real—at least at our within-brain, neuronal level—than had been thought. When we see a horror movie, part of the horror evidently comes from the fact that on some level we are literally experiencing it, just as an X-rated movie can arouse sexual feelings among those watching—all of which suggests some wisdom in the Buddhist insistence that we should distinguish "mental formations" from the real thing.

Psychoanalysts might wonder, as well, whether mirror neurons provide the basis for transference and countertransference, if neural circuits associated in an analysand's mind with, say, her father are activated by the experience of talking with her analyst.

[iv] The psychological term for assessing the subjective mental processes of another individual.

Mirror neurons also suggest a hardwired neuronal basis for compassion and empathy tied to our own evolutionary needs. Maybe autism, in turn, results from faulty mirror neurons or an insufficient number thereof; ditto—albeit with important differences—for sociopathy, since sociopaths notably lack empathy. Do practicing Buddhists and Buddhist masters in particular have especially acute or abundant mirror neurons? They certainly demonstrate unusual amounts of compassion.

Of late there has been considerable research on the neurobiology of empathy, an inquiry likely to experience substantial growth in the immediate future. For example, a recent and comprehensive review by psychologists from Harvard and Columbia Universities, titled "The Neuroscience of Empathy: Progress, Pitfalls and Promise," took stock of the state of the field, concluding that three basic systems are involved.[7] First is "experience sharing," whereby another's emotions are felt as though they are one's own; thus, another's happiness can make one happy, his or her sadness makes one sad, and so forth. The second is "mentalizing," or intellectually and consciously assessing these sensations and using cognitive powers to make sense of the situation. Finally comes "prosocial concern," whereby observers are motivated to reach out to those in pain. The result is a hierarchy of involvement, beginning with shared feeling, then feeling combined with understanding, and, lastly, feeling combined with both understanding and the sense of a need to act, to become engaged, on behalf of those who are suffering.

Buddhists typically extend this triumvirate of empathic concern beyond other humans to "all sentient beings," with appreciation of karma typically adding some additional intellectual motive that helps to fuel the transition from shared feeling to engaged action. This progression, from passive identification to active participation in alleviating *dukkha*, might not require an assist from mirror neurons, but they can't hurt.

As a statement of the linkage between cause and effect, karma illuminates what we might call the universality of radical contingency: the extent to which things and actions depend on other things and actions. As such, it certainly isn't foreign to Buddhism. An awareness of radical contingency was key to the Buddha's own intellectual awakening: "One who sees conditioned arising sees the dharma," he explained in one of his earliest teachings. "And one who sees the dharma sees conditioned arising." All happenings are fluid, structured by what precedes, and as a result of this dependence, whatever has happened, is happening, and will happen is not foreordained. Since it need not have occurred, it could occur differently, so it depends—among other things—on our karma: how we act.

Speaking to a wandering seeker, the Buddha is reported to have advised, "Let be the past, let be the future. I shall teach you the Dharma: When this exists, that comes to be; with the arising of this, that arises. When this does not exist, that does not come to be; with the cessation of this, that ceases too." Everything that constitutes the present is thus the fruit of the past, and like all fruits it carries within itself seeds of the future—although nothing is guaranteed. As the paleontologist Stephen Jay Gould noted so memorably:

> We are the offspring of history, and must establish our own paths in this most diverse and interesting of conceivable universes—one indifferent to our suffering, and therefore offering us maximal freedom to thrive, or to fail, in our own chosen way.[8]

Karma is essentially the key to freedom. This may seem paradoxical, since by at least one divergent interpretation, karma is a limitation on freedom, insofar as everything is connected to everything else, leading to a kind of billiard ball determinism whereby we could never do anything other than what we have done: it is or was our karma. But the Buddha emphasized over and over that he taught the *dharma* specifically for the sake of actually enhancing freedom so that people could liberate themselves from cycles of ignorance, hatred, and fear.

Earlier I quoted from the novel *Cloud Atlas*, criticizing its naive acceptance of the literal existence of souls. Here is another observation from *Cloud Atlas*, this time the movie version.[v] It begins as follows: "Our lives are not our own. We are bound to others, past and present." Thus far, freedom might seem to be denied, although the acknowledgment of *pratitya-samutpada* is also clear. But the observation goes on, and through it we can glimpse considerable wisdom: "by each crime, and every kindness, we birth our future." Saying it more Buddhistically—and with no less biological accuracy—our bodies and our minds are the direct result of the accumulated actions of immense numbers of beings in the past, and by our actions we influence how things will be in the future. We are karma. We all have great power. And to quote that modern icon of morality, Spiderman, "With great power comes great responsibility."

Responsibility to what? To the world generally, and more specifically to those other sentient beings with whom we are indissolubly linked. Although, as already noted, Buddhism embeds humanity in nature, it also makes a distinction in that people alone are thought capable of working toward their own

---

[v] So far as I can tell, it is not present in the novel.

enlightenment—and thus, somehow, toward the enlightenment of all other sentient beings—via a conscious choice to do so. It's our karmic opportunity as well as our burden.

It is all too tempting to live within our small selves ("me" and "you") while at the same time destroying the very life of our large selves: not just the human "we" and "us" and "them," but also the rain forests, the oceans, the atmosphere—all those "things" that make up part of our larger selves, and upon which we depend.

The issue is potentially richer and deeper than it first appears. What about viewing karma as essentially the way we construct our lives—not in some misty, mystical future existence occupying the bodies of animals or other people-to-be, but right here and now, in the only existence we definitely know? The Christian pacifist A. J. Muste famously wrote that there is no way to peace—peace is the way. Centuries earlier, the Jewish philosopher Baruch Spinoza, in the final proposition of his *Ethics*, concurred with Aristotle's observation in the *Nicomachean Ethics* (written a mere century or so after the Buddha lived) that happiness—*eudaimonia*, in Greek—isn't our reward for living a virtuous life; it is virtue itself.

Conversely, sin isn't something that we are punished for, but rather something that we suffer by. We become the lives we lead, regardless of whether such living leads in turn to any consequential life events the next time around. Karma, I am suggesting, isn't something we carry like an albatross about our necks so much as it is something that we construct as we go, inseparable from life itself, thereby creating obligations and opportunities that are equally inseparable. As the Buddhist scholar William Waldron points out, the very concept of karma carries with it a deep human responsibility, worthy of Spiderman and then some:

> by clearing away the false dichotomies and useless debates over unchanging natures, entities and essences, Buddhist and biological views on the groundlessness of life highlight an immediate vision of ethical responsibility—*just because* it emphasizes our radically constructed and interdependent nature. If we end up becoming what we do, we live *out*, not just *with*, the consequences of our actions.[9]

Recall the Parable of the Poisoned Arrow, in which the Buddha emphasized the importance of avoiding sterile philosophical debates over the intricacies of doctrine and instead concerning ourselves with life as it is, here and now. Consider, as well, the sense in which Buddhism has earned its designation as the Middle Way. The Jains of India have long been particularly concerned about not harming other living beings, an attitude that ranks high among Buddhists as well, but generally within practical limits. Some Jains have carried their sense of *ahimsa* (non-harming) so far as to starve themselves rather than eat. By contrast, part of the immersion of Buddhism in biology is its practical Middle Way recognition that human beings are inextricably part of the living world, such that we literally cannot survive *without* the taking of life, even if it is the life of carrots and coconuts.

The Buddha is said to have elected a middle way between severe, self-injurious asceticism and gluttonous debauchery. There is, therefore, in Buddhism a down-to-earth recognition of reality and of the circumstances that on occasion require realistic departures from the ideal. Here, accordingly, is the Parable of the Drowning Woman, in which we are again advised to focus on matters of the lived and living world rather than hitching our concerns to sterile doctrine: once upon a time, two monks were walking along a river when they heard the cries of a woman who was drowning. One of the monks dived in and saved her, while the other looked on with disapproval. Later, after they returned to their temple, the disapproving monk berated the other for having abandoned one of the primary rules of their *sangha* (the local community of Buddhist devotees), which prohibits touching the body of a woman. The other monk replied, "I let go of the woman when I brought her to the riverbank. You are still clinging to her!"

For engaged Buddhists, the lesson is clear: the world is drowning, and there is no shame in trying to rescue it. The law of karma may even require that we do so. In addition, a recognition of *anitya* leads to the acknowledgment that conscious intervention carries with it the genuine prospect of making things better, if only because the *anitya*-infused nature of reality means that change is not only possible, it is inevitable. The question thus isn't will there be change? but what sort of change will it be? (And of course, the doctrines of *anatman* and *pratitya-samutpada* mean that in helping to rescue the world, things are made better for one's greater self as well.)

In the same manuscript in which Stephen Batchelor proposed Buddhism "version 2.0," he explains that his endorsement of secular Buddhism makes use of the original Latin derivation of the word "secular." Its roots lie in the word *saeculum*, which, as Batchelor points out, means "this age, this *siècle* (century), this generation." Batchelor therefore describes his plea for secular Buddhism as based upon "those concerns we have about this world, that is everything that has to do with the quality of our personal, social, and environmental experience of living on this planet."[10]

I cannot think of a more appropriate basis for establishing a system of philosophical and ethical actions, whether or not they are accompanied by traditional religious beliefs.

Unlike the Abrahamic big three, which simply proclaim *what* we should do—because God commands it—Buddhism (whether secular or not) actually explains *why*, and it does so without invoking divine edict. Rather, it appeals to a fundamentally naturalistic statement of the way things are—call it the *dharma*, or, equivalently, a deep appreciation and understanding of the natural world. Fyodor Dostoevsky, in *The Brothers Karamazov*, famously announced (through the mouth of Ivan, one of the brothers), that without God, all things are possible—meaning that there is no alternative basis for morality. This is widely considered a problem for atheists and is something to which Buddhism offers a coherent response: fundamental morality (loving-kindness, generosity, caring, etc.) are mandated not by the arbitrary commands of God, a particular god, or an array of gods, but by the karmic structure of existence itself.

Part of Buddhism's engagement with nature is motivated by the pleasures of ego expansion, which, without genuine paradox, is not distinguishable from ego demolition.[vi] There are some other interesting aspects of karma. For one, a useful purpose is served even by misunderstanding it as a kind of Santa Claus list of the extent to which we have been naughty or nice. Karma is thus a practical goad to be good, to overcome our otherwise natural, evolved inclination toward selfishness. Evolution by natural selection is nothing if not selfish, hence the phrase made famous by Richard Dawkins: "the selfish gene."[vii]

Buddhist ethics therefore interact with biology in a nontrivial manner by encouraging followers to act counter to their more immediate, instinctive tendencies toward selfishness. In this sense, there is likely a practical, prosocial advantage in the widespread interpretation of Buddhist karma as a kind of "what goes around comes around" payback for misdeeds.[viii] Worrying about their karma has doubtless kept many a Buddhist on the straight and narrow, a consequence not especially different from the purported benefit of Christian and Islamic teaching about saving one's soul by obeying the commands of God, which often requires forgoing certain troublesome, self-serving inclinations such as stealing, murdering, committing adultery, and so forth.

The mainstream interpretation of karma can also have a liberating effect—precisely the opposite of fatalistically justifying social inequity, the misbehavior of others, or the random reality of simple rotten luck. Thus, karma seen as responsibility for one's actions is notably empowering, telling us that our lives are ours to construct,

---

[vi] The more people understand that their seemingly private and personal egos are meaningless and literally impossible without the interpenetration of everything else, the larger those egos become, until they are so large that they don't qualify as small, individual ones.

[vii] Note that this does not imply that genes are conscious, intentional actors. Rather, those genes that through their effect on the bodies in which they reside are successful in projecting more copies of themselves are necessarily favored—i.e., positively selected—relative to alternative genes whose impact is the furtherance of comparatively fewer self-copies.

[viii] If the current generation continues—as seems likely—to abuse the planetary environment, notably via anthropogenic global warming, succeeding generations will suffer as a result of our karmic misbehavior.

and as a result, many Buddhists—especially in modern times—give great weight to intentionality and human responsibility.

Of course, Buddhism and biology do not coincide and mutually reinforce each other in all respects. One of the intriguing disagreements derives from the Buddhist embrace of free will, which conflicts with modern biology, specifically neurobiology and its subjectively awkward but scientifically necessary denial of free will.[ix] The Western empirically based argument is, in essence, that free will is impossible insofar as all behavior (including thoughts, perceptions, and sensations) results from brain activity, which in turn is due entirely to physical, chemical, and electric events whereby ions enter and leave neurons in precise response to concentration gradients, electrochemical charges, and other factors. As a result, strict adherence to a mechanistic, materialist view of brain function (and there is no other scientifically valid alternative) leaves us with the necessary conclusion that free will is an illusion, even though our subjective experience cries out that it is ever present. In this respect, Buddhism is on the side of common sense rather than science.

Buddhism isn't alone in this regard. William James, who became the first professor of psychology in the United States, wrote in 1890 that the "sting and excitement" of human life derive from our sense that "things are really being decided from one moment to another, and that it is not the dull rattling off of a chain that was forged innumerable ages ago." Paradoxically, this dull, rattling chain that James so denigrated is depicted in Buddhist tradition as unavoidable karma, which has indeed been "forged innumerable ages ago."[11] But as we'll see in the next chapter, it is possible to embrace both karma and free will; Buddhists, in fact, find it impossible to avoid doing so.

We have noted that Buddhists pursue their efforts toward enlightenment "for the good of all sentient beings." But how would it benefit a polar bear whose habitat is melting if a Buddhist meditator in Nepal, or Staten Island, achieves peace of mind? For Buddhists espousing a strict view of karma and reincarnation, such benefit exists because we are literally each other's ancestors as well as descendants. Alternatively, maybe this benefit arises simply because it enhances the total amount of good karma

---

[ix] It is said that when Isaac Bashevis Singer was asked whether he believed in free will, he replied "Of course! I have no choice!" A cute story, this, initially because of its immediate paradox: if Singer has no choice in the matter, then this contradicts his supposed belief in free will. But there is also another, deeper paradox here: we are forced to believe in free will, or rather to act as though we do, in order to live our lives. But science demands otherwise.

mystically floating around in the world, a small but not inconsequential drop of beneficence in a competitive and often uncaring totality.

"To advance compassion and yet survive in a world of appetites," wrote the marine biologist Carl Safina, sounding much like a Buddha, "that is our challenge."[12] Of course, there are related challenges, including this rather cynical one, not typically raised, but which I cannot help considering: maybe Buddhists—especially monks who have taken a vow of poverty and dependence upon public generosity—have long promoted the idea that their spiritual practice benefits everyone and everything as a selfish way of encouraging the laity to donate. One of the rules of the Buddhist road is that monks aren't supposed to store food or labor for it; rather, they are obliged to survive via alms donated by others. It may be, therefore, that these others are encouraged to participate by the overt claim that by doing so they are achieving "merit," which will lighten their karmic burden. If so, this is not too different from the medieval Catholic practice of selling indulgences, which so incensed Martin Luther.

For now, we must leave unresolved the question of whether early Buddhist teaching was essentially self-serving when it emphasized the value of acquiring merit via generosity toward Buddhist functionaries themselves. There can be no doubt, however, that there is also a long-standing and definitely *unselfish* Buddhist tradition that derives the principles of non-harming (*ahimsa*) and loving-kindness (*metta*) from *anatman*, *anitya*, and *pratitya-samutpada*, and that urges karmic activity toward the natural world in a way that can legitimately be described as biologically benevolent and eco-friendly for its own sake.

The Sigalovada Sutra urges, for example, that a person should gather wealth like a honeybee collects pollen: noninjuriously, and thus without destroying, depleting, or in any way diminishing the flower. In the Karaniyametta Sutra we are advised to have reverential loving-kindness toward all creatures, big and small, tiny and great, visible and invisible, near and far, long and short, timid and bold, those already born and those awaiting birth. And Santideva, whom we encountered in chapter 4, encapsulated much of Buddhism's activist pronaturism, foreshadowing the engaged Buddhism of today:

> For as long as space endures
> And for as long as living beings remain,
> Until then may I too abide
> To dispel the misery of the world.[13]

Dispelling the misery of the world, liberating all beings, treating other life forms as delicately as a bee treats a flower: these are tall orders but, given the basic orientation of Buddhism and its parallels in modern biology, neither surprising nor unrealistic. Walt Whitman was neither a Buddhist nor a biologist, yet when he wrote that "whoever walks a furlong without sympathy walks to his own funeral in his shroud," he had an imaginative foot in biology and Buddhism alike, as did Albert Schweitzer, an ardent Christian, when he urged that "until he extends the circle of his compassion to all living things, man will not himself find peace."

This second chapter on engaged Buddhism might well conclude with yet another parable, in which two brothers were urged to search for a treasure buried in their father's garden. They found neither gold nor silver, but their digging greatly enriched the soil.

7

Meaning (Existential Bio-Buddhism?)

AT ONE POINT in Douglas Adams's hilarious science-fiction fantasy *The Hitchhiker's Guide to the Galaxy*, a sperm whale plaintively wonders to himself: "Why am I here. What is my purpose in life?" as he plummets toward the fictional planet Magrathea. This appealing yet doomed creature sought to know its "meaning," but had none. Moments before, it had simply been "called into existence" several miles above the Magrathean surface when a nuclear missile, directed at our heroes' space ship, was inexplicably transformed into this particular organism via an "Infinite Improbability Generator." (A second missile was turned into a pot of petunias, but that is another story.)[i]

Evolution, too, is an improbability generator, although its outcomes are considerably more finite. In this concluding chapter, I shall argue that, having been called into existence by the particular improbability generator called natural selection and then inseparably, irretrievably, and yet impermanently immersed in the rest of life, none of us has any more inherent purpose than Adams's naive and ill-fated whale, whose blubber was soon to bespatter the Magrathean landscape, or the elk you met in chapter 4, who—even as you read this—is still decomposing nicely, yet meaninglessly.

This chapter is something of a departure from the preceding ones in that I'll be presenting my personal perspective rather than attempting to describe objectively the interface between Buddhism and biology. (Some critics—Buddhists as well as

---

[i] As it happens, the petunias said, "Oh, no, not again!"—which we might take as the author's subtle bow to Nietzsche's "myth of the eternal return"...or not.

biologists—will doubtless complain that I've been developing my own idiosyncratic viewpoint on both Buddhism and biology all along, so that the present chapter differs only in being more honest about its subjectivity.) In any event, it is impossible to consider "the meaning of life" without being at least somewhat personal and subjective. So here goes.

From a narrowly biological perspective, we have one and only one purpose in life, although, strictly speaking, it isn't our purpose but rather that of our genes: to transmit as many of themselves as possible into the future. Our biological purpose, as organisms put together by natural selection working on those genes and gene combinations, is to maximally facilitate this transmission. For most of us, this isn't likely to make the heart sing, perhaps because human beings—alone, it appears, among living things—are capable of stepping back and evaluating evolution's strategy. We and, I suspect, we alone are able to look deeply into those tactics employed by our genes, even though they created us and did so entirely for their benefit, not ours. Moreover, in an increasingly exploited, overpopulated, and underresourced world, there is little to be said in favor of this particular biological goal and much that commends opposition.

Biology, in short, offers us a meaning, but one that simply isn't very satisfying and that may even be downright hurtful.

A traditional Buddhist perspective offers very little about the meaning of life beyond suggesting that our purpose is to become as enlightened as possible, and that we can do so by liberating ourselves from error and freeing ourselves from enchainment by *samsara*—literally, the flow of the world, especially its repeated cycles of birth, death, and rebirth. Eventually, after proceeding through numerous such incarnations (or transmigrations—see chapter 5), particularly adept Buddhist practitioners are ostensibly able to achieve *nirvana*, freedom from the *dukkha* that necessarily accompanies *samsara*.

Ancient Buddhist *sutras* and *tantras* typically don't speak directly about the "meaning" or "purpose" of life, and indeed, I have been unable to identify a suitable Sanskrit or Pali term that can reasonably be translated this way.[ii] Instead, the focus is on the potential to end suffering by removing the causes of suffering, which in turn are identified as various cravings and conceptual misattachments.

The deeply pessimistic worldview of the German philosopher Arthur Schopenhauer (1788–1860) has parallels to Buddhism. Schopenhauer himself acknowledged this and in his later life derived some satisfaction from the

---

[ii] The tantras are a ritualized ancient Indian system of teaching and practices that seeks to capture part of the world's energy via specialized meditative techniques.

convergence, although he clearly stated that his familiarity with Buddhism came only after he had written his great tract *The World as Will and Representation*. In any event, the crux of the Schopenhauer-Buddhism connection is that for Schopenhauer, people are driven by their desires and cravings (for Buddhists, *tanha*), and are made unhappy (*dukkha*) by failure to achieve them combined with an inability to let go of trying. Moreover, there is also the problem that even if these cravings are gratified, they don't remain so for long. Attachments generally—and cravings in particular—may be important to evolutionary success (more on this later), but according to traditional Buddhist thought (notably, the second of the Four Noble Truths), they are inimical to happiness.

Whether you take a narrowly biological perspective—in which your life is merely a kind of genetic catapult—or an equally narrow Buddhist one, in which your life is at best merely a steppingstone to enlightenment, something is missing. Another way of putting it: biology and Buddhism both teach that life has no inherent meaning, that neither you nor I nor anyone else was in any sense "put here" in order to achieve some deeper purpose. We have no basis for criticizing the whale of Magrathea, no matter how foolish he seems.

When it comes to the human search for meaning, my advice, accordingly (which is neither strictly biological nor Buddhist but, I believe, fully consistent with a modern understanding of both traditions) is: stop looking for it, if you interpret "looking for it" as trying to find something that was there all along, intended specifically for you, and simply waiting to be revealed. It is, instead, something to create.

I would go further and suggest that the traditional Christian perspective, according to which we have been put here for a reason—to serve some purpose other than our own—is actually rather degrading and perhaps even offensive compared with one that grants us the opportunity to decide upon the why and wherefore of our lives. I submit, instead, that our human responsibility (and opportunity) is to achieve meaning by how we choose—mindfully, as many Buddhists emphasize—to live, and, moreover, to do so often in direct resistance to our nature, as contained in our genes. The Abrahamic big three typically seek to instill meaning by combinations of commandment, exhortation, and threat, backed up by the promise or peril of an afterlife. By contrast, instead of threatening us with hell, Buddhism—especially in its ancient, traditional form—offers the promise of liberation from supposedly repetitive and punishing reincarnations in future lives. Perhaps the conceptual tradition that is closest to the one I espouse is not religious at all, but rather the philosophical approach known as existentialism. Thus, this chapter's meditation is an elaboration—for the first time, I believe, in the history of the world!—of "existential bio-Buddhism."

In an uncaring universe, but one in which we are deeply embedded, the human project, I maintain, is to define ourselves by our persistent struggle, even if—indeed, especially if—like Sisyphus, we know that we will be unsuccessful.

I want to exhort you, as well, to resist exhortations and to live meaningfully in a world devoid of inherent meaning. I also feel a need to identify my own orientation, which—in addition to being highly sympathetic to a kind of atheistic Buddhism—also resonates especially with the atheistic existentialists, notably Jean-Paul Sartre and Albert Camus (neither of whom, incidentally, was particularly enamored of Buddhism).

Evolutionary science essentially points us toward a perspective whereby all living things—human beings not least—are playing a vast existential roulette game. No one can ever beat the house. There is no option to cash in one's chips and walk away a winner. The only choice is to keep on playing, and indeed, some genes and phyletic lineages manage to stay in the game longer than others. But not forever. Where is the meaning in a game whose goal is simply to keep on playing, that can never be won, only lost, and in which no one got to design the rules?

In short, there is no intrinsic biological meaning to being alive. We simply are, and inter-are. Ditto for our genes. Indeed, we are because of our genes, which are because their antecedents have thus far avoided being eliminated. As Martin Heidegger—a twentieth-century precursor of Sartre—put it (and as the Buddha would doubtless have agreed), we have simply been "thrown into the world." None of us, after all, was consulted beforehand. Our genes did it—or rather, our parents' genes, and their parents' before them. (Karma, anyone?)

At this point, biologists and Buddhists might well join in chorusing: how absurd! How meaningless to have been produced in this way, and for such a self-gratifying purpose—namely, the perpetuation of the genes themselves. Indeed, it isn't really much of a purpose at all, especially because it was never consciously chosen.

Biology tells us about the world as it is—essentially nothing about what we ought to do. Existentialism tells us about what we ought to do—essentially nothing about the world as it is.[iii] And Buddhism tells us a fair amount about both. Put them all together, and we get an intellectual trifecta that tells us quite a lot about everything.[iv]

[iii] To be fair—and I thank James Woelfel for this—existentialism also seeks to inform us about how we perceive the world.

[iv] Personal note: my literary agent suggested that I refrain from introducing existentialism into this book, worrying that it is not only passé but also too highbrow. Although he is very intelligent and wise in the ways of Western publishing, my hope is that you, dear readers, will prove him wrong, at least in this regard.

For one of the more interesting correspondences between existentialism and Buddhism, consider responsibility for one's actions. Both Søren Kierkegaard and Friedrich Nietzsche are considered early founders of existentialism (even though the term wasn't invented until after their deaths) because, among other things, they both strongly emphasized the need for people to take responsibility for their actions; both thinkers—and the existentialists who followed them—maintained that we are essentially nothing (more accurately, that we do ourselves a deep dis-service) until we create ourselves by how we choose to live our lives. The Buddhist concept of karma is almost exactly the same, focusing on how any action has a double effect: directly, with regard to that which is acted upon, and indirectly, by reflecting back upon the actor. Thus, for Buddhism no less than for existential-ism, we are all works in progress who manufacture what we take to be ourselves by what we do.

One glory of Buddhist insight is that it urges us to choose, mindfully and yet with equanimity. For its part, existentialism has been dubbed the philosophy of freedom, since it, too, strongly emphasizes our responsibility to define ourselves by the choices we make. The glory of biology is that it gives us the power to delve beneath the surface of things and thus the opportunity to act meaning-fully upon our choices. The upshot of all this? Decomposing nicely, while also striving for the betterment (Buddhists would say the "liberation") of all sentient beings.

Devotees of oxymorons ("freezer burn," "jumbo shrimp," "military intelligence") might well have a field day with the gravamen of this chapter, which conjoins not two apparent opposites, but three. On the other hand, just as triangles, tripods, and three-legged stools are remarkably stable, existential bio-Buddhism may prove sur-prisingly sturdy, not merely as a concept but also as a practical structure. One caveat, however: there is no perfect correspondence between existentialism, biology, and Buddhism, just as the convergence between biology and Buddhism—as we have seen—isn't exact. Instead, useful insight can be gained by considering both the simi-larities and the differences among all three traditions.

One of Buddhism's many functional parables is the story of pilgrims who come to a rushing river, which they cross by taking advantage of a boat moored nearby. After ferrying themselves to the other side, they are ill-advised to carry the boat with them for the rest of their overland journey. Similarly, we would be wise to allow the useful articulations between existentialism, biology, and Buddhism to carry us as far as we find convenient, but to let go of any discon-nects that aren't useful, so long as we understand where the newly minted boat (perhaps a three-hulled catamaran) serves us well and where we had better pro-ceed on foot.

First, let's turn to some of the more prominent incompatibilities between existentialism and biology. One of existentialism's organizing principles is the basic notion that human beings have no "essence." In that regard it joins hands with Buddhism's embrace of not-self, or *anatman*. As Simone de Beauvoir wrote, a person is "the being whose essence is having no essence" (*l'être donc l'être est de n'être pas*). Or, as Sartre famously put it, "existence precedes essence." For the existentialists, there is no Platonic form of the person, no ideal essence—basically, no *atman* à la Hindu, Christian, or Muslim thought—of which our corporeal reality (our physical existence) is but a pale instantiation. Rather, we define ourselves, give ourselves meaning, establish our essence only via our existence: by what we do, by how we choose to live our lives. All this because, once again, we have no indelible, internal kernel of human nature independent of the specifics of how each of us chooses to live.

The concept of choice turns out to be especially important here, because in the view of the existentialists, we are free; indeed, in Sartre's paradoxical words, we are "condemned to be free." Lacking any inherent self-substance other than our own freedom, we are forced to make choices, and in doing so we define ourselves. In a huge universe, devoid of purpose and uncaring about people, it is our job to give meaning to our lives by the free, conscious, intentional choices we make. So far, so Buddhistic.

There is a difference, moreover, between this existentialist conception that there is no human essence and that presented to us by biology. Thus, at the heart of a purely evolution-based conception of human nature—or of hippo, halibut, or hickory tree nature—is the recognition that living things are skin-encapsulated congeries of genes, jousting with other, similar genes (more precisely, alternative alleles) to get ahead. Free, conscious, intentional choices seem out of place for any creature that is merely the physical manifestation of genes preprogrammed to succeed.

Here is a brief selection from the *Letters* of John Keats, written about two hundred years ago, long before sociobiology and even before Darwin. Yet it captures much of the essence (if you'll pardon that word) of modern evolutionary thinking:

> I go amongst the fields and catch a glimpse of a stoat or a fieldmouse peeping out of the withered grass. The creature hath a purpose and its eyes are bright with it. I go amongst the buildings of a city and I see a man hurrying along—to what? The creature hath a purpose and its eyes are bright with it.

For evolutionary biologists, too, living things have a purpose, one that is shared by stoats, field mice, and human beings. What is that purpose? It is, as already noted, the projection of their genes into future generations. Or, as Richard Dawkins and others have emphasized, the behavior of living things can be viewed as resulting from the efforts of their constituent genes to get themselves projected into the future. Living things are survival vehicles for their potentially immortal genes. Let me repeat: biologically speaking, this is what they are, and *all* that they are.

Most existentialists can be expected to disagree. What about Buddhists?

Buddhism, not surprisingly, is silent on the role of genes, since they were only identified with the work of Gregor Mendel in the nineteenth century, confirmed by numerous subsequent studies of animal and plant heredity, and finally structurally decoded by the Watson-Crick molecular model in the mid-twentieth century. But although most Buddhists—ancient as well as modern—can be expected to approve the existentialist's notion that we lack a permanent, independent essence, they would dispute the caricature that we are nothing but helpless marionettes, doomed to dance at the end of our respective DNA strings. As noted earlier, Buddhist thought makes much of the role of free will in addition to a notably modern, deep sense of responsibility, summarized by karma.

It is important to observe in this regard that molecular biology long ago rejected the idea that genes determine outcomes—whether anatomic, physiologic, or behavioral—with anything approaching rigid control. There are numerous genes, for example, whose sole function is to regulate the activities of other genes, and the surrounding environment itself modifies gene expression in crucial ways. Our genes whisper to us; they do not bark orders. Thus, Buddhism and existentialism are closely allied when it comes to the question of free will in that both acknowledge its presence and, moreover, they both celebrate it. By contrast, a strictly biological mind-set, insofar as it is materialist, balks at the very idea—not so much from its focus upon genes as because of its commitment to material causation.

This is because, as discussed in the last chapter, if the mind derives entirely from physical actions in the realm of neurobiology—and so far as we can tell, it does—then thoughts, feelings, and conscious actions must also be the consequence of charged ions crossing nerve cell membranes; and such a naturalistic, automatic process leaves no room for free will. Or, as Schopenhauer put it (without benefit of neurobiology), "A human can very well do what he wants, but cannot will what he wants."

The only scientifically valid alternative to materialist causation would be a literally uncaused, spontaneous event, such as the behavior of a radioactive nucleus when it unpredictably throws off alpha or beta particles, or gamma rays. But insofar as such events are truly random and spontaneous—and one could argue that nothing really is—the result is hardly bedrock for free will. Alternatively, if neurobiological phenomena are physically caused after all, then free will once again must be abandoned.

Even though such abandonment accords quite closely with a strictly scientific worldview, it goes against the widespread commonsense perspective by which each of us feels that he or she is fundamentally in control of our thoughts and actions—even if not quite sovereign when it comes to emotions. No less a scientist than Albert Einstein actually derived comfort from the assumption that people aren't necessarily responsible for their actions, especially when these actions are regrettable. "This knowledge of the non-freedom of the will," he explained in a 1932 speech given to the German League of Human Rights, "protects me from losing my good humor and taking much too seriously myself and my fellow humans as acting and judging individuals."[1]

Here, then, in the realm of free will, we have a case in which existentialism and Buddhism join forces in opposition to a strictly anti–free will, biologically confirmed viewpoint, in the process sharing a perspective that, although unscientific, is also one that accords very well with nearly everyone's subjective experience. It is difficult indeed to find anyone who isn't privately convinced that he or she has free will.[v]

Thus, Buddhist thought diverges from materialist biological science in asserting that genuine intentionality exists even though strict cause-and-effect thinking (supported by biology) insists that free will is an illusion.

A Buddhist mind-set, incidentally, leads to conclusions at odds with biology when it comes to another important natural manifestation as well: mutations. These are considered by biologists to be random copying errors that occur more frequently in certain genes than in others. But just as physicists cannot predict exactly when a given nucleus of an unstable radioactive isotope will throw off ionizing radiation, thereby becoming more stable, molecular biologists cannot predict exactly when a given gene will mutate—although they can say with considerable confidence that given a large population of genes, accurate statistical predictions can be made as to what proportion of these genes will have mutated in a specified time, just as physicists can do with respect to a sample consisting of many radioactive nuclei.

The Dalai Lama, interestingly, despite all his support for science, is quite uncomfortable with randomness generally, and particularly as it is manifested in mutations. "From a philosophical point of view," he wrote,

> the idea that these mutations, which have such far-reaching implications, take place naturally is unproblematic, but that they are purely random strikes me as unsatisfying. It leaves open the question of whether this randomness is best

---

[v] There is yet another problem with the Buddhist embrace of free will, one that I cannot solve but nonetheless feel obliged to acknowledge: how to reconcile *anatman*, *anitya*, and especially *pratitya-samutpada* with free will? Given the realities of not-self, impermanence, and interpenetration, isn't "freedom" unavoidably constrained?

understood as an objective feature of reality or better understood as indicating some kind of hidden causality.[2]

This highly influential Buddhist is so convinced as to the universality of a materialist, cause-and-effect-dominated universe that he gravitates strongly toward anything— even something as seemingly unscientific as "hidden causality"—that could make mutations "caused" rather than random. It remains unknown at present whether any comparable hidden causality will ever be found to underpin free will. (I'm betting against it.)

For the evolutionary biologist, behavior is one way genes go about promoting themselves. (Other ways are by producing a body that is durable, adapted to its eco-logical situation, and capable of various physiological feats such as growth, metabo-lism, repair, etc.) This is what behavior *is*. And, biologically speaking, that's all that it is. It certainly seems that here—contrary to both the existentialist and the Buddhist position—is a conception of a human essence, one that is exactly coterminous with human DNA. Moreover, our essence—our genotype—seems in this formulation to *precede* our existence, exactly contrary to what Jean-Paul Sartre and other existen-tialists would have us believe. Moreover, there doesn't appear to be much room for meaningful freedom of choice—so beloved of existentialists and Buddhists—insofar as we are slaves to the selfish genes that created us, body and mind.

But stoats and field mice, hippos and hemlock trees don't know what they're doing, or why. Human beings do. Or at least, they know what they are doing when-ever they mindfully assert themselves, refusing to allow themselves to be pushed and pulled this way and that by the evolutionary whisperings of their DNA.

"Man is a thinking reed, the weakest to be found in nature," wrote the French mathematical genius, religious mystic, and precursor of existentialism Blaise Pascal:

> But he is a thinking reed. It is not necessary for the whole of nature to take up arms to crush him: a puff of smoke, a drop of water, is enough to kill him. But, even if the universe should crush him, man would still be more noble than that which destroys him, because he knows that he dies and he realizes the advantage which the universe possesses over him. The universe knows nothing of this.[3]

Thanks to biological insights, people are acquiring a new knowledge: what their own genes are up to, what their evolutionary purpose is. But this need not necessarily be their existential purpose. Instead, I argue that an important benefit of evolutionary wisdom is that by giving us the kind of knowledge about the universe that Pascal so admired, biology leaves us free at last to pursue our own chosen purposes, whatever they may be—and that engaged Buddhism helps point toward such a purpose.

Pascal prefigured existential thought in other respects as well, as when he wrote that "the silence of these infinite spaces frightens me." Such fear is understandable, since the comfortable sense of human specialness that characterized the pre-Copernican world was being replaced in Pascal's day (the seventeenth century) with a vast universe of astronomical distances, no longer centered on *Homo sapiens*. The great, empty spaces of evolutionary time and possibility—as well as the kinship with "lower" life-forms that it demands—has frightened and repelled many observers of evolutionary biology as well. "Descended from monkeys?" the wife of the bishop of Worcester is reputed to have exclaimed. "My gracious, let us hope it isn't true. But if it is true, let us hope it doesn't become widely known!"

Well, it *is* true, and it is becoming widely known. But like Pascal and the existentialists who followed him, our place in the biological universe remains for many a chilling reality. "When you look into the void," wrote Nietzsche (a more immediate ancestor of existentialism), "the void looks into you."

Nietzsche adopted as one of his more notable exhortations a sentence from the classical Greek poet Pindar: "Become what you are." Nietzsche took this to mean something like "move beyond the illusory sense of your ideal self and accept your true existence as a collection of instinctive drives." In other words, embrace yourself as you are rather than what you have been persuaded to think that you ought to be. In short: *amor fati*—"love your fate." Once you can identify your true self, urges Nietzsche, embrace it and don't let anyone, anything, or any social constraints make you untrue to it. Buddhists similarly emphasize the importance of seeing through delusion (*maya*) and perceiving the nature of your true self, although they maintain, unlike Nietzsche, that this truth is not founded on individual separateness but on a profound lack of separateness, a recognition of *anatman* and *pratitya-samutpada*—as well as the fact that owing to *anitya*, whatever passes for your self is always changing.

Both existentialism and modern evolutionary biology focus on the smallest possible unit of analysis: either the self or the gene. The Danish philosopher and existential pioneer, Kierkegaard, asked that this only should be written on his gravestone: "The Individual." And in his masterful *Man in the Modern Age*, the psychiatrist and philosopher Karl Jaspers dilated upon the struggle of individuals to achieve an authentic life in the face of pressures for mass conformity. For Buddhists, by contrast, individuals are illusory, and our smallest unit is actually quite large, encompassing the rest of the living world (remember Indra's net).

The individual and gene-centered perspective of biology has given rise to criticism that when applied to human behavior, it is inherently cynical, promoting a gloomy, egocentric worldview. The same, of course, has been said of existentialism, whose stereotypical practitioner is the anguished, angst-ridden loner, wearing a black turtleneck and obsessing, Hamletlike, about the meaninglessness of life.

Add Buddhism to the mix and we have a somewhat different perception: that on the one hand (at the level of strict evolutionary functionality) each of us is best understood as the product of distinct genes coming together to establish every "individual," while on the other (an equally valid, contemporaneous perspective) our "true selves" can be grasped only by appreciating our not-self (*anatman*), impermanence (*anitya*), and interconnectedness (*pratitya-samutpada*).

Let's grant (if only for now) that human beings—like other living things—are merely survival machines for their genes, lumbering quasi-robots whose biologically mandated purpose is neither more nor less than the promulgation of those genes. If so, then there is no more inherent meaning to life as seen in evolutionary terms than when viewed by the existentialists. For most biologists, the promulgation of genes is neither good nor bad. It merely is, just as, for Buddhist modernists, we just are, or, rather, we inter-are. Although scholars (and some scoundrels) have occasionally attempted to derive ethical guidelines from evolution, so-called evolutionary ethics has not fared well. This is because such formulations run afoul of what the philosopher G. E. Moore has labeled the "naturalistic fallacy," first elaborated in slightly different form by David Hume: the erroneous expectation that "is implies ought."

Although it is tempting to conclude that the natural world provides a model of how human beings ought to behave, it does no such thing. This is clearly seen, for example, when we examine such perfectly natural entities as the virus that causes AIDS: what could be more "organic" than this, made of protein and nucleic acids? Yet, with the exception of those fundamentalist nutcases such as Pat Robertson who claim that AIDS is God's righteous response to sexual sin, no one would characterize HIV as "good." In the previous chapter, we considered many other examples of how the natural, biological world leads unavoidably to *dukkha*.

Those deluded by the promise of evolutionary ethics are probably confusing a laudable bias toward things that are "natural" and "organic" (organic foods, untrammeled wilderness, natural childbirth, etc.) with the sense that *anything* natural or organic must be inherently desirable. Most evolutionary biologists, by contrast, know that the organic world—although fascinating—is neither pleasant nor moral. Once again, it just is. Indeed, some biologists, notably the eminent theoretician George C. Williams, have emphasized that if we must judge evolution in ethical terms, then if anything, it is downright bad: cruel, selfish, shortsighted, indifferent to the suffering of others, and so on. The fruits of evolution, just like the process itself, may inspire our admiration for their complexity and subtlety, but not for anything approaching saintliness or even benevolence. They are certainly not "role models."

It is well-known that existentialists are very much occupied with the inherent meaninglessness of life and the consequent need for people to assert their own meaning, to define themselves against an absurd universe that dictates that ultimately everything will come to naught, because they will die. Less well-known—but, I believe, equally valid—is the fact that whereas evolutionary biology makes no claim that it or its productions are inherently good, it—like existentialism—teaches that life is truly absurd.

Poets, to be sure, have looked for the "meaning of life," and some—in my opinion, the most honest—have acknowledged its absence, perhaps none more forthrightly than Heinrich Heine in his poem *Questions*, which tells of a man who asked the waves, "What is the meaning of Man? Whence did he come? / Whither does he go? / Who dwells up there on the golden stairs?" In response, "the waves murmur their eternal murmur, / the wind blows, the clouds fly, / the stars twinkle, indifferent and cold, / and a fool waits for an answer."

We must conclude that there is no answer, no intrinsic meaning to the brute biological fact of being alive. We simply are. And ditto for our genes. Indeed, we are because of our genes, which are because their antecedents have avoided being eliminated—such that they constitute our karma. Since each organism—notwithstanding *anatman, anitya,* and *pratitya samutpada*—is a unique concatenation of molecules as well as past experiences, each has a unique karma. But this does not in itself imply that each is accordingly meaningful.

The Buddhist modernist perspective—particularly exemplified, as we have seen, by the teachings of Thich Nhat Hanh—would add that it isn't quite true that we simply are; rather, we complexly inter-are. We have indeed been thrown into the world, but instead of worrying about why (remember the Buddha's Parable of the Poisoned Arrow), we have the opportunity to know our true selves—that is, our interconnectedness as well as our impermanence and lack of self—and mindfully make of it what we will. In other words, we have the unique opportunity to infuse our lives with meaning, in proportion, ironically, as we recognize that each personal existence isn't narrowly "one's own life" alone!

Some might say at this point that if evolutionary biology reveals that life is without intrinsic meaning, then this simply demonstrates that biology is mistaken. Not at all. From the perspective of natural science generally, there is no inherent reason why anything—a rock, a waterfall, a horsefly, or a human being—is in itself meaningful. Certainly this is what the existentialists have long emphasized, pointing out that the key to life's meaning is not aliveness itself, but what we attach to it. Kierkegaard, for example, felt passionately about the need for people to make their lives meaningful, just as the Buddha taught that his Eightfold Way offers a path to enlightenment.

Among existentialists in particular, the sense of self-construction is far from accidental, because for them there is no reason to suppose that meaning comes prepackaged along with life itself, even human life. Kierkegaard once wrote about a man who was so abstract and absent-minded that he didn't even realize he existed, until one day he woke up and found that he was dead! And Buddhist tracts—ancient as well as modern—are filled with stories about the need for people to silence their chattering "monkey minds" and be "in touch with the world" as it exists "in the present moment."[vi]

People are nonetheless left with a biological legacy that inclines us—if left to our own devices—to behave in concert with our selfish genes rather than with our wider, kinder, more Buddhistic selves. In addition to the more obvious and typically troubling tendency for selfishness, thoughtlessness, excessive competitiveness, and even, on occasion, violence, human beings harbor other, milder but no less misleading inclinations. For example, natural selection is likely responsible for something that Buddhists in particular have recognized and regretted: the fleeting nature of happiness and satisfaction. Unfortunately for human contentment, it is biologically adaptive for cravings (Sanskrit *tanha*) to occur and then recur. Imagine an animal who, after eating or copulating just once, says to itself, "Well, that takes care of that"! Biological needs regenerate, and to see that those needs get acted upon, happiness and satisfactions have evolved to be fleeting.

The Buddha seems to have understood this, so that whereas the Middle Way counsels that certain needs ought to be fulfilled, we are most human when we learn to say no to others, especially those that we mistake for fundamentals but are merely *maya* (illusion). An important parable suggests that it can be useful to point at the moon, but important not to mistake our finger for it.

On the one hand, biology and Buddhism are contradictory, since evolution inclines us to be selfish whereas Buddhism teaches compassion. There is room, nonetheless, for convergence: biological selfishness is often achieved by altruism toward kin, toward reciprocating partners, and so on. But Buddhism also departs from biology in teaching compassion toward all. It is a near-constant struggle to make sense of truly selfless compassion; in this respect, Buddhist teachings run counter to evolutionary inclinations. I could go further and suggest that in some cases, Buddhist teachings are difficult precisely *because* they work against millions of years of evolution. But that does not mean that we are stuck with an atomized, un-Buddhist mind-set, since to a large extent, minds are among those things that we are free to make up for ourselves...admittedly with a gentle shove from that array of genetically inspired tendencies that we designate "biology."

[vi] Another personal note: my wife and I have named our camper *The Present Moment*, so that when we are away on a hiking trip, we can honestly claim to be "in *The Present Moment*."

There are in fact many ways that human beings can and do say no to their genes, just as Sartre and Camus, for example (and Kierkegaard and Nietzsche before them), encouraged people to rebel in their personal lives. We may elect intentional childlessness. We may choose to be less selfish and more genuinely altruistic than our genes might like. We may decide to be "genuine Christians," à la Kierkegaard, and/or refrain from hurtful speech, vocation, intentions, and so forth. In short, we may elect to follow the Eightfold Way, à la Buddhism. A perspective suffused with existential bio-Buddhism suggests that in fact we *must* do these sorts of things if we want to be fully human. The alternative—to let biology carry us where it will—is to forgo the responsibility of our humanity, and to be as helpless and abandoned as that (briefly) airborne Magrathean whale we considered some time ago, or those mired in *maya*, *dukkha*, and *tanha* whom the Buddha sought to liberate.

Part of existential "bad faith"—as initially developed by Sartre and elaborated by others afterward—is a continuing pattern of delusory flight from one's own freedom and the responsibility it entails. In this regard there is both similarity to and difference from Buddhism: although both Buddhists and existentialists revere the idea of human freedom, they differ in that whereas the existentialists focus on the self as the unit of such freedom, Buddhists emphasize the absence of an independent self (*anatman*). This, in turn, leads to an important problem that, I must confess, I am unable to solve: how can Buddhism espouse freedom without also identifying the self as the locus of that freedom? Can any individual be truly free, given that he or she is actually part of a larger whole? Might it perhaps be held that submersion in a larger whole actually facilitates greater "individual" freedom—indeed, that it constitutes the ground substance from which the greatest possible freedom emerges? Maybe. Or maybe not. (As the king of Siam acknowledges in the movie *The King and I*: "Is a puzzlement!")

But perhaps this is a case of worrying about the source of that Poisoned Arrow....

In any event, "going with the flow" of our biologically generated inclinations is very close to existential "bad faith," wherein people pretend—to themselves and others—that they are not free, whereas they in fact are. This is not to claim that human beings are perfectly free. When José Ortega y Gasset observed, for instance, that "man has no nature, only a history," he neglected that this includes an evolutionary history, as the result of which we are constrained in some ways as well as impelled (or, rather, predisposed) in others. We cannot assume the lifestyle of honeybees, for example, or Portuguese men-of-war. But such restrictions are trivial and beside the point, which is that within a remarkable range, our evolutionary bequeathal is almost wildly permissive—more so, it appears, than that of any other animal.

It is interesting, in this regard, to consider how much time and energy people expend trying to induce others, especially the young and impressionable, to practice what is widely seen as the cardinal virtue: obedience. To recast Freud's argument about incest restraints: if we were naturally obedient, we probably wouldn't need so much urging to do what we are told. And yet, on balance, I would like to propose that far more harm has been done throughout human history by obedience than by disobedience, and I therefore wish to suggest the heretical and admittedly paradoxical notion—based on existential bio-Buddhism—that in fact we would be well advised to teach more disobedience. Not only disobedience to political and social authority, but especially disobedience to some of our own troublesome genetic inclinations.

In short, I reject the notion that we should knuckle under to that tyranny purported to emanate from our genes, just as I reject the widespread Judeo-Christian-Islamic claim that we are put here on earth to contribute to or in some other way work toward God's plan—whether for his greater glory, or our salvation or anyone else's. I accept, instead, a very limited version of the strictly evolutionary, biological perspective which holds that we are here to maximize our inclusive fitness: we are indeed here because of the biological process of evolution, since, after all, it was the success of our parents' genes, and that of their parents', and so on, that produced us. But the fact that we were produced by natural selection (and that genetic fitness maximization is therefore our karma) doesn't mean that we are in any way obliged—either biologically, Buddhistically, or ethically—to go along with its promptings.

Sartre would doubtless agree, since he argued in unintended parallel with Buddhism that each person (a being *pour-soi*, or "for-itself"), as opposed to the unconscious external world (objects *en-soi*, or "in-itself"), is empty of independent existence and must therefore create its unique selfhood. And for existentialists, we are all free to do as we choose. Indeed—and this may well be key—for Sartre it is because we are fundamentally undefined, lacking in an essential self, that we are not only free to choose but even, paradoxically, "condemned to be free." Like it or not, we have to choose. For Buddhists, by contrast, freedom arises when we realize our impermanence and achieve nonattachment. But it's still a choice.

The fact that human beings are influenced in various ways by their genes—the subject matter of sociobiology or evolutionary psychology—is, I believe, terribly important, and well worth our study and understanding. As I have already suggested, maybe such understanding is even necessary in order for us to explore the potentials of our own freedom. If this seems incongruous, bear in mind that genetic influence is a far cry from genetic determinism. There is very little in the human behavioral repertoire that is under genetic control, yet very little that is not somehow affected by our genes. At the same time, human beings are remarkably adroit at overcoming such influences as exist.

For a change of pace, consider the game of volleyball. Talented volleyball players do extraordinary things, making amazing leaps and spectacular saves to keep the ball from touching the ground. Their game can be seen as a metaphor for much of human life. The evolutionary imperatives of projecting our genes into the future—technically, maximizing our "inclusive fitness"—is like the action of gravity on a volleyball. It works, persistently—even remorselessly—in a certain direction. Without substantial efforts on our part, if we stop diving and leaping and batting the ball into the air, gravity wins. But people are incredibly adroit at keeping it airborne. We have invented all sorts of cultural rules, social mores, and systems of learning and tradition, some of which support biology and many of which contravene it.

We may not literally define our essence by our existence, as the mid-twentieth-century existentialists proclaimed, but a deep understanding of sociobiology suggests that the existentialists and Buddhists were absolutely right: when it comes to the most important and interesting aspects of our behavior, genes make suggestions; they do not issue orders. It is our job, our responsibility, to choose whether to obey. We are free, terrifyingly free, to make these decisions, to keep the ball in the air.

Volleyball is a particularly useful metaphor since it also implies a team effort, and human beings are notably social creatures. In his play *No Exit*, Sartre observed that "hell is other people," since they interfere with one's freedom, but it remains true that human beings generally do not—and cannot—avoid other human beings. Whether dilemma or delight, our social relationships are also very much the stuff of exciting and important evolutionary insights, illuminating our cooperation with kin and reciprocators as well as our competition with others.

At the same time, the volleyball image may be troublesome—although no less accurate for that—insofar as it also implies competition: after all, the reason the ball is kept off the ground is so that it can be smashed successfully onto the opponent's court. Nonetheless, for teammates, it is a non-zero-sum game,[vii] in which the coparticipants share the prospect of enjoying win-win outcomes. Compared to their zero-sum-game alternatives, non-zero-sum games lend themselves to enhanced compassion and mutual kindness; they generate positive inclinations, facilitated by interdependence (remember *pratitya-samutpada*?).

Instead of volleyball, let's try tennis, a game in which the players are in a zero-sum relationship with whoever is on the other side of the net, the "opponent." I put this word in quotes because in a real sense, two tennis players aren't opponents at all but coparticipants, without whom no pleasurable game can occur. By contrast, there is

---

[vii] In zero-sum games, a winner necessarily implies a loser, since the total at the game's end is always zero: a winner and a loser. Non-zero-sum games, by contrast, can terminate with lose-lose or win-win outcomes. Not surprisingly, the latter are to be preferred.

no need for verbal niceties when it comes to playing doubles: you and your tennis partner are clearly in a non-zero relationship: she wins, you win; she loses, you lose.

It can be argued that in the real (e.g., biological) world, zero-sum games do in fact occur, as for example in the case of predator versus prey. But even here, there is a non-zero component, although one not immediately apparent. Each literally depends on the other, such that the tennis metaphor breaks down—or at least, the idea of opponent(s). One can say that in a deep sense, we are all playing doubles and without a net—or, insofar as there is one, we are all on the same side of it. Time now to introduce another metaphor, this one from existentialism: the myth of Sisyphus. You may recall that Sisyphus, a figure from Greek mythology, was an especially tricky fellow who had irritated the gods and was punished by having to spend eternity pushing a heavy rock up a steep hill, only to have it roll back down again. Sisyphus's punishment, therefore, lasted forever, because his task was never completed. When Albert Camus retold this story in a famous essay, he emphasized that Sisyphus possessed a kind of nobility precisely because he knew that his efforts were in vain. Sisyphus will never succeed, just as we will never succeed in living forever, in winning the all-night, all-day, all-time poker game, in contravening our biology. And yet Sisyphus perseveres. He has identified his task, his job in life, all the while knowing its hopelessness and its absurdity. He struggles on anyhow, because that is what it means to be a fulfilled human being. Camus ends his essay by going even further, with the stunning announcement that Sisyphus is *happy*.

Camus is the existentialist thinker most associated with the "life is absurd" characterization of the human condition. Often misunderstood, he felt that this absurdity didn't reside in life itself, but in something uniquely human, namely the relationship (which he called a "divorce") between the human need for ultimate meaning and the "unreasonable silence" of the world. Neither human existence nor the universe is absurd in itself; rather, it is the relationship between the two, whereby people seek something of the universe that it fails to deliver.[viii]

I would like to suggest that there are some similarities between the volleyball of human biology, desperately kept above the ground by our various cultural stratagems, and the rock of Sisyphus. In the end, the game is hopeless; biology wins (each of us eventually dies), and the ball hits the ground, just as the rock rolls downhill. Moreover, as we have seen, the actual game—like the task of Sisyphus—is absurd, meaningless in itself. Purveyors of early Buddhism also perceived the absurdity and responded by suggesting an array of paths—eight of them, to be precise—whereby every Sisyphus could be spared his repetitive and hopeless task. Buddhist modernism, as we have seen, counsels a somewhat different route, in which we mindfully

---

[viii] I thank James Woelfel of the University of Kansas for clarifying this for me.

and even joyously put our combined shoulders to the rock, while if possible also seeing ourselves as not fundamentally different from it.

Perhaps our existential bio-Buddhistic purpose is to make our lives meaningful by emulating Sisyphus. And perhaps here is yet another way that evolution comes in: just as, for the existentialist Camus, Sisyphus achieves a kind of grandeur because he struggles on, fully aware that for him success is literally impossible—that is, knowing with certainty that the world is stacked against him as an individual—I am not alone in suggesting that evolution offers us a kind of potential grandeur that exceeds the solitary splendor of existential heroism and goes along with that of Buddhist *anatman*, *anitya*, and *pratitya-samutpada*. Here is the final paragraph of *The Origin of Species*:

> It is interesting to contemplate a tangled bank, clothed with many plants of many kinds, with birds singing on the bushes, with various insects flitting about, and with worms crawling through the damp earth, and to reflect that these elaborately constructed forms, so different from each other, and dependent upon each other in so complex a manner, have all been produced by laws acting around us.... Thus, from the war of nature, from famine and death, the most exalted object which we are capable of conceiving, namely, the production of the higher animals, directly follows. There is grandeur in this view of life ... that, whilst this planet has gone cycling on according to the fixed law of gravity, from so simple a beginning endless forms most beautiful and most wonderful have been, and are being evolved.

Evolution, especially with its sociobiologic updating, offers us a terrifying insight, a degree of self-knowledge: an understanding of what it is that our genes are up to. Then it leaves us on our own to decide whether we shall sit back passively or struggle against this attempted tyranny, like Sisyphus, with all our strength. Add Buddhism and existentialism to the mix, and we have a profoundly liberating perspective by which we are not only empowered with a measure of free will, but also encouraged how to engage our lives: on behalf of the rest of life, to which we are inextricably connected.

But where might the necessary strength come from, and of what does it consist? In the *Laws*, Plato's last and longest dialogue, Socrates comments that we are like marionettes, with the gods pulling our strings at their whim. At the same time, he points out, we have one golden string available by which to pull back, to assert ourselves in return. He was referring to our use of reason, by which, according to Socrates, Plato, and generations of Western thinkers ever since, we are to reclaim our identities, our independence, our unique status as autonomous entities.

And yet, if you are fond of ironies, here is a good one: the history of human thought—that is, employing this special golden string by which human beings are enabled to distinguish themselves—has led to a progressive debunking of humanity's sense of its own specialness, including even its vaunted rationality. Intellectual history has been, in a sense, an ongoing series of earthquakes, which add up in various ways to a continuing onslaught upon the proposition that *Homo sapiens* is uniquely wonderful. The irony, of course, is that our capacity for complex thought (our ability to pull back against the gods with this golden string) is itself one of our most remarkable—and special—qualities; and yet, at the same time, it is responsible for a diminution of our specieswide claim to unique status on Earth, if not in the cosmos.

Thus, with the demolition of the Ptolemaic, Earth-centered universe, our planetary home was relegated from centrality to periphery, and *Homo sapiens*, by implication, along with it. But at least *we* remained, self-designated as the apple of God's eye, made in his image. Such a conceit became difficult to maintain, however, with the identification of evolution itself (followed a century later by sociobiology's ongoing elucidation of the behavioral implications of natural selection and its gene-based orientation). Then came Freud's discovery of the unconscious and the sobering fact that we are not even masters in our own house. Existentialism, too, has been touched by these various intellectual earthquakes, notably Nietzsche's hyper-Darwinian insistence that in a values-free world, human beings must rise above traditional morality and define themselves as *Übermenschen*. In both a Darwinian and an existential sense, even as our species becomes less central, each individual becomes more.

Here Buddhists will dissent, since, as we have seen, for them individuality is itself largely a misconception, even though human beings are granted pride of place in most Buddhist cosmologies. At the same time, however, a deep sense of *anatman*, *anitya*, and *pratitya-samutpada* militates against going too far along the existential path of personal loneliness.

Evolutionary science tells us that we are the physical results of past struggles to maintain bodies and especially to pass along genes. Interactions between our ancestors and their environments that led to greater success led to more copies of those genes responsible for that greater success, but not necessarily to anything good, if by "good" we mean anything other than brute biological success per se.[ix] Nor would it necessarily lead to happiness, although to some extent we might expect this to be so,

---

[ix] Bear in mind that "brute biological success" simply refers to the bottom-line consequence of genes promoting copies of themselves into the future; it does not necessarily imply that brutes are especially successful.

since happiness is the body's way of telling us that we are experiencing something likely to contribute to that success. (Probably "pleasure" would be more appropriate, emphasizing its typically short duration.)

The Buddhist lexicon, mostly Sanskrit based, contains many terms describing those factors that contribute to *dukkha*. Thus, *tanha* refers to striving or craving, and can, in turn, be seen as deriving from two root sources: *raga*, or desire generally, and *dosa*, which is its polar opposite, aversion. Translated into biological reality, *raga* equates to the urge to obtain mates, food, living space, stuff, and so on, and *dosa* is fundamentally defense (of self, family, territory, etc.) and also aggression against whatever may be threatening. A key point for our purposes is that these are all behaviors that have been encouraged by natural selection as means to maximizing fitness. Whether or not they are ethically good, they are certainly natural; thus, our understanding of biology helps us understand the underlying basis of what Buddhists consider to be pain and suffering.

Similarly, although a strong case can be made (indeed, I hope the previous chapters have gone some way toward doing so) that "selfhood," as typically understood by most non-Buddhists, is in fact misleading, an equally strong case can be made that the actual perception of selfhood is itself a product of evolution by natural selection. In order for those genes that temporarily help constitute an individual body to be evolutionary successes, it is necessary that their bodily possessors behave in a way that assumes that those bodies are in fact rigidly distinct and separate from other bodies.[x] There is, incidentally, no conflict between the underlying fact that individuals really aren't distinct and separate, and the other fact that when it comes to genes and bodies maximizing their biological well-being, there is a payoff in being, in a sense, deceived. This is because there is no reason for evolution necessarily to promote accuracy as such; rather, it favors any tactics and strategies that generate success, regardless of whether this is achieved by brains that have an accurate perception of the world or by brains that are mired in error.

Most of the time, of course, there is a distinct biological payoff to perceptual accuracy: which plants to eat and which to avoid, how best to approach a wounded mastodon or a potential mate, and so forth. But "scientific" accuracy per se is not privileged over error so long as the misperception has no practical, negative consequences. For example, living things do not suffer from their perception that "solid" objects are truly solid, even though modern (and more accurate) physics tells us that

---

[x] An important exception occurs when individuals treat other individuals—their genetic relatives—as though they are proportionally more or less identical with the behaving individual. Not surprisingly, such treatment involves behaving benevolently toward others in direct proportion as those others are genetically close relatives—which is to say, in proportion as the bodies in question are in a real sense extensions of one another.

they are mostly composed of empty space. It is entirely likely that the widespread but uninformed impression that individuals in most cases are *not* interconnected has itself been promoted by natural selection. It is therefore a responsibility of biologists and Buddhists to help correct this understandable but troublesome error.

Similarly, there are other cases in which natural selection has promoted behavioral tendencies that are adaptive but—even as they help generate evolutionary success— that also lend themselves to generating unhappiness, especially when combined with cultural processes that cater to these basic (and often base) inclinations. It is a goal of modern Western psychology, for example, to help people become "well adjusted." Or, as Freud put it, "Where id was, there shall ego be." This isn't the Buddhist goal, although Buddhist thought has certainly struggled with the causes of human unhappiness. Although Buddhism is if anything more distrustful of the id than was Freud, Buddhist psychology—and there definitely is such a thing—also has relatively little sympathy for the ego, focusing instead on ways to transcend faulty perceptions of the ego's needs (even its existence) and searching for the causes of egoistic dysfunction.

For example, Buddhism identifies "three poisons" (*trivisa*) that are especially responsible for *dukkha*. These are greed (the grasping, never satisfied desire for *more*), ill-will (including hatred as well as anger), and delusion (or ignorance). These unwholesome roots of personal distress have parallels at the larger level of societies as well. Thus, greed is a driving force behind rampant dog-eat-dog capitalism, consumerism, indifference to social injustice, rapacious treatment of natural ecosystems, and so forth, whereas ill-will on a societal scale results in militarism as well as harsh retributive justice. And delusion, which is bad enough when experienced by individuals, is readily enhanced by the mass media.

At the personal level, susceptibility to each of the three poisons has unfortunately been bequeathed us by evolution, with the impact of these poisons often amplified by modern life. Greed, for example, involves a fondness for possessions, something that in moderation contributes to fitness. But since evolutionary fitness is relative to that of others, rather than absolute, we have not been outfitted with a shut-off valve once a certain threshold of possessions is achieved. As a result, people are often stuck in a never-ending cycle of attempting to meet a "need" that cannot be satisfied. Attachments of one sort or another—especially to material objects—may thus be important to evolutionary success, but inimical to happiness.

Hatred—which Buddhists view with repugnance—likely derives from a similarly adaptive tendency, this time toward competition (especially male-male, at least in its more violent aspects), ramifying into family versus family and even group versus group. Ignorance is a bit more complex, but it might relate, among other things, to a number of tendencies that are (or were) adaptive in preindustrial times but are troublesome today. First is the already-mentioned delusion of being an isolated and permanent self; it is likely that any of our ancestors who lacked this perspective would not have ended up being our ancestors.

Something may thus be true—such as the fundamental interconnectedness of living things, especially members of the same species—but nonetheless lead to diminished biological fitness. Although it may contribute to genetic success, a sense of being separate and disconnected also contributes to widespread alienation and thus considerable mental anguish, even as it may be promoted by natural selection.

In addition, we typically give greater credence to negative experiences than to positive ones, since it pays, for example, to take the threat of a saber-toothed cat more seriously than the pleasure of a blue sky on a sunny day. And don't forget the widespread, deep discomfort that most people have with transience—notably, death. It should be emphasized once again that natural selection doesn't care one bit whether we are happy or suffering, only that we project our genes into the future, and the regrettable reality is that evolution has endowed us with these and other poisons—not really for our own good, but for that of our genes. Unsatisfied, unfriendly, and deluded people may well be considerably less happy than if they were enlightened, but their ignorance could nonetheless have been selected for.

It is even possible—perhaps likely—that our big brains have also caused more than their share of trouble. There is little question that intelligence was positively selected during human evolution; the evidence is undeniable in the rapid increase in brain size from prehominids to *Homo sapiens*. And although the precise adaptive factors that drove this evolutionary change are unclear, they likely involved some combination of enhanced communication, tool use, more effective exploitation of the local environment, perhaps a well-honed ability to associate positively with some while competing with (and/or deceiving) others, and so forth.[4] Regardless of which factors helped generate our braininess, it also seems undeniable that along with an advanced intellect has also come some serious problems, foremost among which is probably our cognitively constructed machinery of destruction, notably nuclear weapons but not excluding anthropogenic climate change, habitat destruction, shortsighted resource exploitation, and many others, including, ironically, advances in public health and hygiene, which in turn have led to potentially disastrous overpopulation. A cartoon appeared some years ago set in a future world that had been turned into a smoldering ash heap by a thermonuclear holocaust. Deep in the oceanic abyss, a ramshackle bunch of exiguous surviving amoeboid creatures have just held a protozoan town hall meeting at which they grimly decide to try again to repopulate the Earth; but before they depart, they take a solemn vow: "This time, no brains!" There may well be other, less dire *dukkha*-generating consequences of our big brains, including susceptibility to depression, paralyzing self-consciousness, and the chattering "monkey-mind" that Buddhists attempt to quiet via meditation.

Thoreau—not coincidentally, an early admirer of Buddhist thought—wrote movingly about his own struggles to quiet his mind and to embrace instead the immediacy of natural experience. This is from his essay "Walking":

Of course it is of no use to direct our steps to the woods, if they do not carry us thither. I am alarmed when it happens that I have walked a mile into the woods bodily, without getting there in spirit. In my afternoon walk I would fain forget all my morning occupations and my obligations to society. But it sometimes happens that I cannot easily shake off the village. The thought of some work will run in my head and I am not where my body is—I am out of my senses. In my walks I would fain return to my senses. What business have I in the woods, if I am thinking of something out of the woods?

It is likely that throughout much of our long prehistoric walkabout childhood on the Pleistocene savannahs of Africa, our ancestors didn't have much time or opportunity to sit around kvetching or berating their overactive minds. Similarly, worry about meaning and meaninglessness arises only when we have enough free time to support such activities, when we feel free to walk in the woods for emotional sustenance rather than to grub for mushrooms.

To be sure, the brain does many other things. It works beautifully, for instance, as a simulator of experiences, thereby sparing us the need to deal with negative outcomes that can be imagined, rather than experienced, and thereby avoided. At the same time, doing so pulls us "out of the moment," which in turn produces the kind of alienation that Buddhist teaching encourages us to overcome and that existentialism seeks to confront. Even empathy and compassion—which are not only Buddhistic virtues but also, in many cases, biological benefits as well—unavoidably generate their own load of pain. In proportion as you really feel another's pain, then, whatever else is going on, you are stuck with, well, feeling that much more pain.

We've seen that evolution does not lead inevitably to an accurate perception of the world. In fact, the opposite is more likely the case: natural selection favors images of self and others that result in maximum fitness, but, paradoxically, this may well involve substantial misrepresentation of the truth, including self-deception as to one's own motives and even the nature of the self.[5] This includes denial of dependence, a deep discomfort with instability and impermanence, and a desire for perpetuation. The illusion of stability, permanence, and a self-in-control may have evolved because they were adaptive, not because they were necessarily accurate or true.

Natural selection, it helps to remember, has acted upon our brains to produce certain results that serve our DNA well but that don't have to reflect the true nature of the world. Thus, the goal of evolution isn't to approach truth but simply to promote genes, and to use any means toward that end. Sometimes these tactics run roughshod over the world as it actually is. For example, we have the illusion that the earth is flat, that the sun moves across the sky, and also—the grand illusion that Buddhists are especially inclined to confront—that each of us is separate from the rest of the world.

The likelihood, however, is that we have been selected to pay special attention to these erroneous perceptions, notably our separateness rather than our connectedness, just as we are predisposed to pay more attention to indications of danger than of ease, relaxation, and peace. If we're wrong in the latter case, the worst that happens is that we fail to take advantage of an opportunity to slow down and smell the roses; if we're wrong in the former, we are liable to end up dead. We are similarly inclined, I suspect, to perceive solidity and stasis, but in the face of reality that is inchoate and dynamic. By the same token, we insist on separateness and boundaries, and on stability, instead of accepting fluidity. Also—and perhaps most troubling—we have been selected to desire and crave—to seek certain situations and conditions and to strive desperately to avoid others, all in the service of maximizing fitness. But while doing so, we subject ourselves and those around us to the downside of such craving, namely suffering, for ourselves and others.

And of course, Buddhism initially gained traction as a route whereby people might transcend this suffering.

Thus, in many cases, biology presents us with a double-edged sword whereby our genetically influenced inclinations contribute to success, but at substantial cost in *dukkha*. Success may be achieved in terms of fitness, while costs accumulate in terms of unhappiness or immorality. In short, adaptedness—crucial for biological success—often comes with a load of woe, not least that classic Buddhist awareness of disease, old age, and death, with the baggage of suffering this entails. We have also considered some of the cravings with which natural selection has saddled us, to which we should probably add a paradoxical craving on the part of many Buddhists: to be released from craving itself.

Despite our complex load of *dukkha*—almost certainly greater than that belaboring any other organism—*Homo sapiens* is also probably the only life-form with the freedom to evaluate its situation. Not surprisingly, therefore, the human search for meaning has been as persistent as it is inchoate. Often it takes the form of religious faith, which can lead in turn to the claim of "nonoverlapping magisteria" or similar nonsense. Yet the reality is that whenever traditional religion, especially in its Abrahamic manifestations (belief without evidence, along with dogma), has infringed on science (conviction based on evidence, along with rational theory), the former has ultimately been forced to retreat. Some argue that a final, secure outpost for religion will be precisely the realm of meaning, the claim that whereas science can tell us what happens, only religion can tell us why. But this is patently false.

Science indeed tells us why things happen: because of thermodynamic, electro-magnetic, or gravitational forces; selection pressure; drifting continental plates; and so forth—including, in many cases, a hefty dose of chaos. It also tells us that—much as many people would like it to be otherwise—the conjunction of certain types of matter known as sperm and egg, nucleotides, proteins, carbohydrates, and a very large number of other physical entities generates human beings, with nothing approaching "purpose" anywhere to be seen.

Moreover, a long-range view of humanity—one that also includes everything else on planet Earth—is hardly encouraging, and not just for the future of *Homo sapiens*. In a notable essay titled "A Freeman's Worship" (1903), Bertrand Russell, one of our most accomplished philosophers and mathematicians, writes that

> All the labors of the ages, the devotion, all the inspiration, all the noonday brightness of human genius, are destined to extinction in the vast death of the solar system, and the whole temple of man's achievement must inevitably be buried beneath the debris of a universe in ruins.[6]

Taking a somewhat less cosmically cataclysmic perspective, personal death, as some existentialists insist on pointing out, makes life absurd. The only thing more absurd would be to deny this absurdity, to be stuck in a meaningless life, recogniz-ing the meaninglessness only dimly, if at all, pretending that Mommy, or Daddy, or Jehovah, or Allah, or Brahma, or the Buddha has everything planned out just for us. Buddhists also take note of the reality of death (along with its concomitants, old age and disease), adding to it the absurdity of endless, meaningless recycling between birth, life, and death.

Back to the cheery whale of Magrathea, with whom this chapter began. On his way down, our smiling cetacean marvels at the large flat thing approaching him very, very quickly. Among his last thoughts: "I think I'll call it 'ground.' I wonder if it will be my friend."

In the right hands, there can be genuine humor in life's meaninglessness. Of the now-vast "literature of the absurd"—much of it theater—a large percentage is down-right funny: black humor, to be sure, but humor nonetheless—including, most notably, the work of Samuel Beckett. As Beckett's novel *Murphy* draws to its antic conclusion, we are given a memorable account of what ultimately became of the hero's ashes, and thus of human goals and aspirations more generally (it is said that Beckett's mentor, James Joyce, was so fond of this passage that he committed it to memory):

> Some hours later Cooper took the packet of ash from his pocket, where earlier in the evening he had put it for greater security, and threw it angrily at a man who had given him great offence. It bounced, burst, off the wall on to the floor, where at once it became the object of much dribbling, passing, trapping, shooting, punching, heading and even some recognition from the gentleman's code. By closing time the body, mind and soul of Murphy were freely distributed over the floor of the saloon; and before another dayspring greened the earth had been swept away with the sand, the beer, the butts, the glass, the matches, the spits, the vomit.

Beckett's best-known work, *Waiting for Godot*, is constructed of similarly shocking and occasionally brilliant absurdities. Like *Murphy*—and like evolutionary biologists with a "gene's-eye" perspective—it confronts the reality of meaninglessness. Two tramps, Vladimir and Estragon, spend all of act 1 waiting for Godot, with whom they believe they have an appointment. He doesn't show up. Nor does he appear in act 2, leading to one wag's description of the play as one in which nothing happens—twice.

Toward the end Vladimir cries to Estragon: "We have kept our appointment, and that's an end to that. We are not saints, but we have kept our appointment. How many people can boast as much?" Vladimir replies gloomily: "Billions."

After all, there is nothing unusual in keeping one's appointment. We all do it, insofar as we exist. Our friend the whale similarly showed up for his appointment with the ground. (Woody Allen once noted that 80 percent of life (or "of success"— it has been reported both ways) is just showing up. Speaking both biologically and Buddhistically, make that 100 percent.) Maybe the key, then, is what you do while you are waiting: for Godot, for a job, a promotion, a mate, a child. Or falling. This is precisely what underlies the literature of existentialism: the prospect of redemption, of achieving meaning via meaningful behavior, even though—or rather, especially because —in the long run any action is meaningless. One of the greatest such accounts of people achieving meaning through their deeds is to be found in Albert Camus's *The Plague*, which describes events in the Algerian city of Oran during a typhoid epidemic.

In this novel, the plague stands for many things, among them the German occupation of France during World War II, the inevitability of death, and, of course, the plague itself. The struggle against cruelty, against death and disease, against an uncaring universe, is testimony to the fact that our creaturely existence is not in itself meaningful. What gives life meaning, therefore, is how one chooses to live. *The Plague* is a chronicle compiled by the heroic Dr. Rieux in order to "bear witness in favor of those plague-stricken people; so that some memorial of the injustice and outrage done them might endure; and to state quite simply what we learn in a time of pestilence: that there are more things to admire in men than to despise."

But Rieux—that is, Camus himself—is quite aware that we always live in a time of pestilence. (Just as the Buddha noted that we always live with *dukkha*.) Camus's essay "The Myth of Sisyphus" makes the point that Sisyphus stands for all humanity. Over and over, generation after generation, death after life after death after life, we push our rocks up a steep hill, only to have them roll back down again; that is our lot, and the lot of our genes.

Camus concludes his essay with the stunning announcement that "one must imagine Sisyphus happy," because he accepts this reality, defining himself—achieving meaning—within its constraints. Camus's stance, in which meaning is not conveyed by life itself but must be imposed upon it, is not only consistent with an informed biological perspective, but actually makes sense only because of that perspective.

As existentialists have long recognized, the quest for meaning itself only takes on meaning—indeed, it only takes place at all—because people are creatures whose lives, embedded in biology, are fundamentally lacking in meaning. If life's purpose were evident, it wouldn't be such a mystery, and existence would be devoid of some of its greatest triumphs, in which human beings struggle to make sense of what is, biologically, a purposeless world: not how people rise above their biology (for that cannot be done), but how they ride their biology to greater heights, or, naively hopeful and enmeshed in *samsara* like the whale of Magrathea, simply plummet to their deaths. Ruminating on Sisyphus, Camus writes that "the struggle itself toward the heights is enough to fill a man's heart." Or as Darwin noted toward the end of *The Origin of Species*, "There is grandeur in this view of life."

At the conclusion of *The Plague*, while the citizens of Oran are celebrating their "deliverance," Dr. Rieux knows better. But at the same time, his commitment to the struggle, to what defines human beings in an otherwise uncaring universe, is undiminished. Rieux understands that in the game of life, all victories are temporary, which renders his perseverance all the more grand indeed:

> He knew that the tale he had to tell could not be one of a final victory. It could be only the record of what had had to be done, and what assuredly would have to be done again in the never ending fight against terror and its relentless onslaughts, despite their personal afflictions, by all who, while unable to be saints but refusing to bow down to pestilence, strive their utmost to be healers.

Healing is precisely what our shared planet needs, what we all need, and it just might be something that a suitable combination of existentialism, biology, and Buddhism can offer. "When I look inside and see that I am nothing," wrote the Indian philosopher Nasargadatta Majaraj, "that's wisdom. When I look outside and see that I am everything, that's love." It may seem saccharine, but perhaps the Beatles were right that "all you need is love."

Nonetheless, critics could well conclude that existential bio-Buddhism is an impossible triple play, and that—to modify slightly the baseball metaphor—in attempting to connect existentialism, biology, and Buddhism, I am guaranteed to strike out. I maintain, by contrast, that the three are natural allies. In italicizing our lack of inherent meaning while striving to respond to our very human, biologically generated need for it, while framing our response in terms of *anatman, anitya,* and *pratitya-samutpada,* Buddhist practitioners are in their own way behaving as existential heroes by refusing to bow down to some of evolution's more unpleasant imperatives and at the same time injecting an appropriately modern, bio-friendly ethic into Buddhism even as we introduce Buddhist concepts into biology. Although no one can announce final victory, I would like to think that Sisyphus—along with Charles Darwin and the Buddha—would applaud such efforts.

Recall Plato's golden string, by which we are granted the strength and opportunity to pull back against the control of the gods. Substitute genes for gods, and you get an oversimplified—indeed, caricatured—evolutionary perspective. Substitute the constraints of society and the inevitability of our death, and you get an oversimplified—indeed, caricatured—existential perspective. Substitute Buddhist compassion and interconnectedness and there is no reason to pull back—indeed, no one to do the pulling!

None of these approaches has settled on a clear alternative to Plato's rope of rationality, although existentialists as well as Zen Buddhists in particular are prone to denigrate the value of "logic chopping," and evolutionary biologists are quick to point out that rationality itself is an adaptation and, as such, situation specific and often blinkered.

Nonetheless, it is important to emphasize that—counter to the Western stereotype of Buddhism as a religion that counsels passive inaction in the face of the world's woes—there exists a vibrant tradition, beginning in fact with the Buddha's own life, that is activist and that urges mindful, rational responsibility for one's actions—maybe even greater responsibility than is found in most Western traditions. For starters, the Buddha is described as having lived a very active life following his enlightenment. Rather than withdrawing into his private nirvana, he went out into the world seeking to share his insights and urging his followers to do the same. Thus, in the *Vinayapatika,* we read the following exhortation:

> I, monks, am freed from all snares, both those of devas and those of men. And you, monks, are freed from all snares, both those of devas and those of men. Go, monks, and wander for the blessing of the many folk, for the happiness of the many folk out of compassion for the world, for the welfare, the blessing, the happiness of devas and men. Let not two (of you) go by one (way). Monks, teach the Dhamma [Sanskrit: *dharma*] which is lovely at the beginning, lovely in the middle, and lovely at the end.

To repeat: Buddhism does not counsel inaction, but rather the avoidance of actions that are inappropriately motivated and generated by cravings and desires that are themselves mistaken. Buddhism also includes, within its toolkit, high regard for "skillful action," or *kusula.*

About 2,500 years later, Jean-Paul Sartre wrote that

> the first principle of existentialism is that man is nothing else but that which he makes of himself.... Thus the first effect of existentialism is that it puts every-man in possession of himself as he is, and places the entire responsibility for his existence squarely upon his own shoulders.[7]

In a sense, traditional Buddhism teaches an even more radical version of this doctrine, whereby not only are we responsible for our current actions, but, according to the doctrine of karma, this responsibility extends into the future and is also reflected in the presumption that the present is a consequence of what "we"—albeit in one or more different bodies—did some time ago. What we are during any given "now" is considered to be a result of what we did at some past "then," and what we do now will impact what we will become.

I, at least, have no difficulty accepting this and indeed embracing it, so long as the focus is on our current, experienced lives. As Cassius famously explains to Brutus in Shakespeare's *Julius Caesar,* "The fault, dear Brutus, is not in our stars but in ourselves." No one is saved or condemned by forces outside him or herself; we are not playthings of the gods (although, admittedly, both random happenstance and the actions of others are consequential in every life). It is ironic, nonetheless, that Buddhism's truly radical perspective of unavoidable responsibility embodied in karma—one that if anything exceeds that of existentialism, which is no slouch in this respect—is married to *anatman,* an equally radical doctrine of not-self.

Even I, an acknowledged Buddhist wannabe and sympathizer, cannot buy the entire package of karmic responsibility when it comes to my supposed past or future existences, or anyone else's. I would love to think that all bad actions are punished and all good ones rewarded, but see no evidence of this (or at least, I see many exceptions, in both directions). Nor can I accept that the effect of my actions will, in any personalized sense, be visited upon me in the form of my subsequent lives after I die. Moreover, even though—to turn to *Julius Caesar* once again—both the evil and the good that we do often live after us, they don't necessarily do so, and certainly there is no evidence that anyone's actions are any more immortal than is each actor.[xi]

In any event, the similarity between Buddhism and existentialism should be clear: both viewpoints make much of people's responsibility for their lives. In *Being and Nothingness*, Sartre italicizes the importance of such responsibility with a statement that could as well have come from the Buddha: "From the instant of my upsurge into being, I carry the weight of the world by myself alone." At the same time, Buddhism isn't unique in offering contradictions. Thus, when he attempts to derive specific moral guidelines for human action (such that, for example, working for social justice is a better choice than working to suppress it), Sartre claims a wider constituency for each behaving individual than that of simply exercising his or her freedom and thereby determining his or her being. "When we say that man chooses for himself," he writes, "we mean that every one of us must choose himself; but by that we also mean that in choosing for himself *he chooses for all men*" (emphasis added).

Maybe Sartre simply means that such choices carry a heavy social responsibility—that is, his claim that when anyone acts he or she is choosing for all people is merely a way of dramatizing the meaningfulness and even the urgency of taking these decisions seriously. (Or maybe—unlikely as it seems—the great Nobel Prize–winning philosopher was just being intellectually sloppy.)

The atheist branch of existentialism—represented by, among others, Beauvoir, Camus, Merleau-Ponty, Nietzsche, Ortega y Gasset, and, most notably, Sartre—was actually outnumbered by the religious proponents, including Berdyaev, Buber, Bultmann, Dostoyevsky, Frankl, Jaspers, Marcel, Pascal, Shestov, Tillich, Unamuno, and most notably, Kierkegaard. That dour Dane asks—actually, demands—something comparable to Sartre's insistence that we are choosing for all mankind: namely, that we behave as though we are acting directly under the gaze of God, hence with "fear and trembling," since, for him, our eternal souls are at stake.

[xi] I am thinking here of the actions themselves, not necessarily their physical manifestations. Obviously, various creations (e.g., buildings, works of art, even offspring) often outlive their creators.

As I mentioned in chapter 6, in *The Brothers Karamazov*, Dostoyevsky—widely regarded as a nineteenth-century precursor of existentialism—maintains through Ivan that in the absence of God, all things are possible, an assertion used as evidence in favor of the value of traditional religion. By contrast, atheistic existentialism (notably expressed by Sartre and Camus) appears to lack any basis for ethical choice, since if each person is truly free to decide on his or her actions, there is no reason to value one decision over another. Sartre sometimes appears to embrace this position, as when, in *Being and Nothingness*, he declares that "all human activities are equivalent...it amounts to the same thing whether one gets drunk alone or is a leader of nations."

And yet, in his later writings, this master of twentieth-century European existentialism comes eerily close to the ancient Buddhist goal of the *bodhisattva*—and of Mahayana Buddhism more generally—maintaining that anyone who values his own freedom is obliged to value and defend the freedom of others, and so must act to further universal human liberation:

> I am obliged to will at the same time as my liberty the liberty of others.... When I recognise...that man is a...free being who cannot, in any circumstances, but will his freedom, at the same time I recognise that I cannot but will the freedom of others.[8]

Both existential traditions—atheist and religious—thus coalesce in emphasizing the enormous responsibility of subjective choice: for Sartre, choosing as though for all humanity, and for Kierkegaard, choosing before God. For traditional Buddhists? Choosing and then acting for karma's sake—given the unavoidable realities of organic interconnectedness; while from biology, silence.

Nonetheless, the prospect remains, fortunately, that human beings—despite their biological baggage—retain enough freedom to fashion their own lives and their own future. "The greatest mystery," according to André Malraux, "is not that we have been flung at random among the profusion of the earth and the galaxy of the stars, but that in this prison we can fashion images of ourselves sufficiently powerful to deny our nothingness."[9] Bertrand Russell is not normally counted among the existentialists, yet his fierce admiration for human freedom justifies his placement among the finest of this breed. Russell, who, as we saw earlier, lamented the ultimate ruin of all things, concluded similarly that the mystery of achieving human meaning in a universe itself devoid of meaning is captured—and to some extent, solved—by the fact that

> Nature, omnipotent but blind, in the revolutions of her secular hurryings through the abysses of space, has brought forth at last a child, subject still to

her power, but gifted with sight, with knowledge of good and evil, with the capacity of judging all the works of his unthinking Mother. In spite of death, the mark and seal of the parental control, Man is yet free, during his brief years, to examine, to criticise, to know, and in imagination to create. To him alone, in the world with which he is acquainted, this freedom belongs; and in this lies his superiority to the resistless forces that control his outward life.[10]

Should anyone doubt the capacity of human beings to deny their nothingness and freely define themselves—if necessary, counter to their evolution-given tendencies—I would like to conclude with a thought experiment that is homey, homely, even scatological, but that should reassure everyone that the biological species *Homo sapiens* possesses abundant room for existential and Buddhistic freedom.

Begin with this question: why are human beings so difficult to toilet train, whereas dogs and cats—demonstrably less intelligent than people by virtually all other criteria—are housebroken so easily?[xii] Take evolution into account and the answer becomes obvious. Dogs and cats evolved in a two-dimensional world, in which it was highly adaptive for them to avoid fouling their dens. Human beings, as primates, evolved in trees, such that the outcome of urination and defecation was not their concern (rather, it was potentially a problem for those poor unfortunates down below). In fact, modern primates to this day are virtually impossible to housebreak.

But does this mean that things are hopeless, that we are the helpless victims of this particular aspect of our human nature? Not at all. I would venture that everyone reading this book is toilet trained! So, despite the fact that it requires going against eons of evolutionary history and a deep-seated primate inclination (or disinclination), human beings are able—given enough teaching and patience—to act in accord with their enlightened self-interest.

For all its mammalian, evolutionary underpinnings, a primate that can be toilet trained reveals a dramatic capacity for freedom, maybe even enough to satisfy the most ardent existentialist and also—not coincidentally—the most committed Buddhist.

An alternative title for this book might have been *Looking Deeply, Seeing Clearly*, because the special charm of identifying connections between biology and Buddhism is less the fascination of *anatman*, *anitya*, and *pratitya-samutpada*, and so

---

xii I apologize to readers who may have encountered this example in my previous writing: some constructs are just too useful to relegate to merely one usage! (In any event, I promise not to do it again.)

forth, than the fact that biology and Buddhism share a rewarding passion for look-ing deeply at the world in general, and at the living world in particular—and, as a result, seeing more clearly.

The struggle to see clearly and deeply, although laudable, isn't easy. Neither is the struggle to infuse our lives with meaning, or simply to know what it is to be human. The writer Thornton Wilder—best known for his play *Our Town*, which did a pretty good job in this respect—once wrote to his sister, who was distraught over reverses she had just experienced in her personal and professional life: "We're all People, before we're anything else. People, even before we're artists. The role of being a Person is sufficient to have lived and died for."[11] Both biology and Buddhism have much to say about being a person, all the more so when the two approaches are employed together, much like the benefit of stereoscopic vision when two eyes focus on the same thing, thereby gaining the perspective of greater depth.

At the conclusion of Ray Bradbury's wonderful science fiction collection *The Martian Chronicles*, there is a startling scene in which a human family, having escaped to Mars to get away from nuclear war, peer together into one of the "canals," hoping to see genuine Martians. They do: themselves. By understanding the convergences between biology and Buddhism, we might all see our genuine selves more clearly and more deeply. Perhaps, moreover, by adding a dollop of existentialism, we might even get some insight into what it is to be a person, and thus, what to live and die for.

Following are the fourteen precepts of the *tien hiep* order of inter-being.[1]

1. *Do not be idolatrous about or bound to any doctrine, theory, or ideology, even Buddhist ones. All systems of thought are guiding means; they are not absolute truth. If you have a gun, you can shoot one, two, three, five people; but if you have an ideology and stick to it, thinking it is the absolute truth, you can kill millions.*

2. *Do not think that the knowledge you presently possess is changeless, absolute truth. Avoid being narrow-minded and bound to present views. Learn and practice non-attachment from views in order to be open to receive others' viewpoints. Truth is found in life and not merely in conceptual knowledge. Be ready to learn throughout your entire life and to observe reality in yourself and in the world at all times.*

3. *Do not force others, including children, by any means whatsoever, to adopt your views, whether by authority, threat, money, propaganda, or even education. However, through compassionate dialogue, help others renounce fanaticism and narrowness.*

4. *Do not avoid contact with suffering or close your eyes before suffering. Do not lose aware-ness of the existence of suffering in the life of the world. Find ways to be with those who are suffering by all means, including personal contact and visits, images, sound. By such means, awaken yourself and others to the reality of suffering in the world.*

5. *Do not accumulate wealth while millions are hungry. Do not take as the aim of your life fame, profit, wealth or sensual pleasure. Live simply and share time, energy and material resources with those who are in need. In the context of our modern society, simple living also means to remain as free as possible from the destructive social and economic machine, and to avoid stress, depression, high blood pressure and other modern diseases.*

6. *Do not maintain anger or hatred. As soon as anger and hatred arise, practice meditation on compassion in order to deeply understand the persons who have caused anger and hatred. Learn to look at other beings with the eyes of compassion.*

7. *Do not lose yourself in dispersion and in your surroundings. Learn to practice breathing in order to regain composure of body and mind, to practice mindfulness, and to develop concentration and understanding. Live in awareness.*

8. *Do not utter words that can create discord and cause the community to break. Make every effort to reconcile and resolve all conflicts, however small.*

9. *Do not say untruthful things for the sake of personal interest or to impress people. Do not utter words that cause division or hatred. Do not spread news that you do not know to be certain. Do not criticize or condemn things that you are not sure of. Always speak truthfully and constructively. Have the courage to speak out about situations of injustice, even when doing so may threaten your own safety.*

10. *Do not use your religious community for personal gain or profit, or transform your community into a political party. A religious community should, however, take a clear stand against oppression and injustice, and should strive to change the situation without engaging in partisan conflicts.*

11. *Do not live with a vocation that is harmful to humans and nature. Do not invest in companies that deprive others of their chance to life. Select a vocation which helps realize your ideal of compassion.*

12. *Do not kill. Do not let others kill. Find whatever means possible to protect life and to prevent war.*

13. *Possess nothing that should belong to others. Respect the property of others but prevent others from enriching themselves from human suffering or the suffering of other beings.*

14. *Do not mistreat your body. Learn to handle it with respect. Do not look on your body as only an instrument. Preserve vital energies (sexual, breath, spirit) for the realization of the Way. Sexual expression should not happen without love and commitment. In sexual relationships be aware of future suffering that may be caused. To preserve the happiness of others, respect the rights and commitments of others. Be fully aware of the responsibility of bringing new lives into the world. Meditate on the world into which you are bringing new beings.*

NOTES

CHAPTER 1

1. Stephen Jay Gould, *Rocks of Ages* (New York: Ballantine, 2002).

2. *A Policy of Kindness*, 67.

3. Stephen Batchelor, *Buddhism without Beliefs* (New York: Riverhead, 1997).

4. David L. McMahan, *The Making of Buddhist Modernism* (New York: Oxford University Press, 2008).

5. Huston Smith, *The World's Religions* (San Francisco: Harper One, 2009).

6. Werner Heisenberg, *Physics and Philosophy* (New York: Harper, 1958).

7. *Kalama Sutta: The Buddha's Charter of Free Inquiry*, translated from the Pali by Soma Thera, is available at http://www.accesstoinsight.org/lib/authors/soma/wheel008.html.

8. Quoted in Tzong Khapa. drang. rages. legs. bahad. snying. po. In *Collected Works of the Buddha*, vol. pha. Delhi: 1979).

9. Matthieu Ricard, *Happiness* (New York: Little, Brown, 2006).

10. Robert A. E. Thurman, *Essential Tibetan Buddhism* (Edison, NJ: Castle Books, 1993).

11. Stephen Batchelor, *Alone with Others* (New York: Grove Press, 1983).

12. George Levine, *Darwin the Writer* (New York: Oxford University Press, 2011).

13. From the *Theragatha* (1062–1071), translated by Andrew Olendzki and published in *Insight* (Spring 1996).

14. Venerable Nanamoli, *The Life of the Buddha* (Kandy: Buddhist Publication Society, 1972).

15. McMahan, *Making of Buddhist Modernism*.

16. Bertrand Russell, *The Scientific Outlook* (London: George Allen & Unwin, 1931).

CHAPTER 2

1. Antonio Damasio, *Descartes' Error* (New York: Penguin, 2005).

2. Albert Einstein, as quoted in *The New York Times,* March 29, 1972.

3. Francis Crick, *The Astonishing Hypothesis* (New York: Scribners, 1995).

4. Antoine Lutz, Lawrence L. Greischar, Nancy B. Rawlings, Matthieu Ricard, and Richard J. Davidson, "Long-term Meditators Self-Induce High-Amplitude Gamma Synchrony during Mental Practice," *Proceedings of the National Academy of Sciences* 101, no. 46 (November 16, 2004): 16369–73.

5. Harold Fromm, "Daniel Dennett and the Brick Wall of Consciousness," *Hudson Review* 59, no. 1 (2006): 161–68.

6. B. Libet, C. A. Gleason, E. W. Wright, and D. K. Pearl, "Time of Conscious Intention to Act in Relation to Onset of Cerebral Activity (Readiness-Potential): The Unconscious Initiation of a Freely Voluntary Act," *Brain* 106, pt. 3 (1983): 623–42.

7. Charles S. Sherrington, *Man on His Nature* (Cambridge: Cambridge University Press, 1942).

8. Ludwig Wittgenstein, *Tractatus Logico-Philosophicus* (New York: Wiley, 1997).

9. David L. McMahan, *The Making of Buddhist Modernism* (New York: Oxford University Press, 2008).

10. Ibid.

11. Thich Nhat Hanh, *The Heart of Understanding* (Berkeley, CA: Parallax Press, 1987).

12. Stephen Batchelor, *Confession of a Buddhist Atheist* (New York: Spiegel & Grau, 2010).

13. The *Vajira Suttra* is available online at http://www.accesstoinsight.org/tipitaka/sn/sn05/sn05.010.bodh.html.

14. David P. Barash, *Revolutionary Biology* (New Brunswick, NJ: Transaction, 2003).

15. Norbert Wiener, *The Human Use of Human Beings* (Boston: Houghton Mifflin, 1950).

16. Sigmund Freud, *On Narcissism*, ed. Joseph Sandler, Ethel Spector Person, and Peter Fonagy (New York: Karmac Books, 2012).

17. Quoted in Rupert Gethin, *The Foundations of Buddhism* (New York: Oxford University Press, 1988).

18. The Dalai Lama, *Essence of the Heart Sutra*, trans. and ed. Geshe Thupten Jinpa (Boston: Wisdom Publications, 2005); emphasis in original.

19. Richard Dawkins, *The Extended Phenotype* (Oxford: Oxford University Press, 1999).

20. Quoted in David Loy, *Money, Sex, War, Karma* (Boston: Wisdom Publications, 2008).

21. Walter Kaufmann, trans., *Nietzsche* (Princeton, NJ: Princeton University Press, 1974).

22. Rupert Gethin, "The Five Khandhas," *Journal of Indian Philosophy* 14(1986): 35–53.

23. Ibid.

24. T. W. Rhys Davids, trans., *The Questions of King Milinda*, Sacred Books of the East 35 (Oxford: Clarendon Press, 1890).

25. Charles Scott Sherrington, *The Integrative Action of the Nervous System* (Cambridge: Cambridge University Press, 1947).

26. Thich Nhat Hanh, *The Sun My Heart* (Berkeley, CA: Parallax Press, 1988).

27. Richard Dawkins, *The Selfish Gene* (New York: Oxford University Press, 1976).

28. Richard Dawkins, *The Extended Phenotype* (New York: Oxford University Press, 1999).

29. Ralph Buchsbaum, *Animals without Backbones* (Chicago: University of Chicago Press, 1987).

CHAPTER 3

1. Thich Nhat Hanh, *The Heart of Understanding* (Berkeley, CA: Parallax Press, 1988).

2. Norbert Wiener, *The Human Use of Human Beings: Cybernetics and Society* (Boston: Houghton Mifflin, 1950).

3. Daniel Kahneman and Jason Riis, "Living, and Thinking about It: Two Perspectives on Life," in *The Science of Well-Being,* ed. Felicia A. Huppert, Nick Baylis, and Barry Keverne (Oxford: Oxford University Press, 2007).

4. Jordi Quoidbach, Daniel T. Gilbert, and Timothy D. Wilson, "The End of History Illusion," *Science* 339, no. 6115 (January 4, 2013): 96–98.

5. Mark V. Barrow Jr., *Nature's Ghosts: Confronting Extinction from the Age of Jefferson to the Age of Ecology* (Chicago: University of Chicago Press, 2009).

6. Francis Bacon, *The Philosophical Works of Francis Bacon*, ed. John M. Robertson (London: Routledge, 2011).

7. Douglas Futuyma, *Evolution*, 2nd ed. (North Scituate, MA: Sinnauer & Associates, 2009).

8. Stephen Batchelor, *Confession of a Buddhist Atheist* (New York: Spiegel & Grau, 2010).

9. Rune Johansson, *The Dynamic Psychology of Early Buddhism*, Institute of Asian Studies Monograph Series 37 (London: Curzon Press, 1979).

10. Robert Aitken, *The Practice of Perfection* (New York: Pantheon, 1994).

11. Jorge Luis Borges, *Labyrinths* (Philadelphia: New Directions, 2007).

12. Ernst Mayr, *Toward a New Philosophy of Biology* (Cambridge, MA: Harvard University Press, 1989).

CHAPTER 4

1. David L. McMahan, *The Making of Buddhist Modernism* (New York: Oxford University Press, 2008).

2. T. Dobzhansky, "Nothing in Biology Makes Sense Except in the Light of Evolution," *The American Biology Teacher* 35 (1973): 125–29.

3. Wm. Theodore de de Bary (ed.), *Sources of East Asian Tradition, Vol. 1: Premodern Asia* (New York: Columbia University Press, 2008).

4. Richard C. Lewontin, "The Organism as the Subject and Object of Evolution," *Scientia* 118, no. 1-8 (1983): 63–82.

5. Richard C. Lewontin, *The Triple Helix* (Cambridge, MA: Harvard University Press, 2000).

6. Eihei Dogen, *Moon in a Dewdrop*, ed. Kazuaki Tanahashi (New York: North Point Press, 1985).

7. Daniel Dennett, *Darwin's Dangerous Idea* (New York: Simon & Schuster, 1995).

8. Thich Nhat Hanh, *Peace Is Every Step* (New York: Bantam, 1991).

9. John Muir, *My First Summer in the Sierra* (New York: Library of America, 1911, 2011).

10. Samuel Taylor Coleridge, *The Friend* (Whitefish, MT: Kessinger, 2004).

11. C. G. Jones, R. S. Ostfeld, M. P. Richard, E. M. Schauber, and J. O. Wolff, "Chain Reactions Linking Acorns to Gypsy Moth Outbreaks and Lyme Disease Risk," *Science* 279 (1998): 1023–26.

12. Hanh, *Peace is Every Step*.

13. Chuang Tzu, *Basic Writings*, trans. Burton Watson (New York: Columbia University Press, 1996).

14. Victor Sogen Hori, "Sweet and Sour Buddhism," *Tricycle*, Fall 1984, 48–52.

15. Stephen Batchelor, *Alone with Others* (New York: Grove Press, 1983).

16. David P. Barash, "Neighbor Recognition in Two 'Solitary' Carnivores: The Raccoon (*Procyon lotor*) and the Red Fox (*Vulpes fulva*)," Science 185, no. 4153 (August 30, 1974): 794–96.

17. Sebastian E. Winter, Parameth Thiennimitr, Maria G. Winter, Brian P. Butler, Douglas L. Huseby, Robert W. Crawford, Joseph M. Russell, Charles L. Bevins, L. Garry Adams, Renée M. Tsolis, Renée M. Tsolis, John R. Roth, and Andreas J. Bäumler, "Gut Inflammation Provides a Respiratory Electron Acceptor for *Salmonella*," Nature 467 (September 23, 2010): 426–29.

18. Edward Abbey, *Down the River* (New York: Plume, 1991).

CHAPTER 5

1. Lynn Townsend White Jr., "The Historical Roots of Our Ecological Crisis," *Science* 155 no. 3767 (1967): 1203–7.

2. Stephen Batchelor, "A Secular Buddhism," *Journal of Global Buddhism* 13 (2012): 87–107.

3. Jerome Kagan, *Three Seductive Ideas* (Cambridge, MA: Harvard University Press, 2000).

4. Shantideva, *A Guide to the Bodhisattva's Way of Life*, trans. A. Wallace and V. Wallace (Ithaca, NY: Snow Lion, 1997).

5. Quoted in David L. McMahan, *The Making of Buddhist Modernism* (New York: Oxford University Press, 2008),

6. Henry Wadsworth Longfellow, *Driftwood* (Charleston, SC: Nabu Press, 2012).

7. Thich Nhat Hanh, "The Individual, Society and Nature," in *The Path of Compassion*, ed. Fred Eppsteiner (Berkeley, CA: Parallax Press, 1985).

8. Dogen (K. Tanahashi and P. Levitt, eds.), *The Essential Dogen—Writings of the Great Zen Master* (Boston: Shambala, 2013).

9. Thich Nhat Hanh, *The Sun My Heart* (Berkeley, CA: Parallax Press, 1988).

10. Thomas Ceary, *The Flower Ornament Scripture* (Boulder, CO: Shambala Publications, 1987).

11. Robert Aitken, *The Dragon Who Never Sleeps* (Berkeley, CA: Parallax Press, 1992).

12. Loren Eiseley, *The Star Thrower* (New York: Random House, 1978).

13. Joanna Macy, *World as Lover, World as Self* (Berkeley, CA: Parallax Press, 1991).

14. Interview for *This Week*, Thames TV, February 5, 1976.

15. Joan Halifax, Foreword to *Buddhist Peacework*, ed. David Chappell (Boston: Wisdom Publications, 1999).

16. Thomas Jefferson, *Jefferson: Writings* (New York: Library of America, 1984).

CHAPTER 6

1. The Dalai Lama, *Essence of the Heart Sutra*, trans. and ed. Geshe Thupten Jinpa (Boston: Wisdom Publications, 2005).

2. David P. Barash, *Revolutionary Biology* (New Brunswick, NJ: Transaction Publishers, 2003).

3. Rune Johansson, *The Dynamic Psychology of Early Buddhism* (London: Curzon Press, 1979).

4. Quoted in W. T. de Bary, ed., *Buddhist Tradition in India, China, and Japan* (New York: Random House, 1969).

5. From Stephanie Kaza and Kenneth Kraft, eds., *Dharma Rain* (Boston: Shambhala Publications, 2000).

6. Evelyne Kohler, Christian Keysers, M. Alessandra Umiltà, Leonardo Fogassi, Vittorio Gallese, and Giacomo Rizzolatti, "Hearing Sounds, Understanding Actions: Action Representation in Mirror Neurons," *Science* 297 (2002): 846–48.

7. Jamil Zaki and Kevin N. Ochsner, "The Neuroscience of Empathy: Progress, Pitfalls and Promise," *Nature Neuroscience* 15, no. 5 (2012): 675–80.

8. Stephen Jay Gould, *Wonderful Life* (New York: Norton, 1989).

9. W. S. Waldron, "Beyond Nature/Nurture: Buddhism and Biology on Interdependence," *Contemporary Buddhism: An Interdisciplinary Journal* 1, no. 2 (2000): 199–226.

10. Stephen Batchelor, "A Secular Buddhism," *Journal of Global Buddhism* 13 (2012): 87–107.

11. William James, *The Principles of Psychology* (1890).

12. Carl Safina, *The View from Lazy Point* (New York: Henry Holt, 2011).

13. Stephen Batchelor, trans., *A Guide to the Bodhisattva's Way of Life* (Dharamsala, India: Library of Tibetan Works and Archives, 1979).

CHAPTER 7

1. Quoted in Michael White and John Gribbin, *Einstein* (New York: Penguin Books, 1994).

2. The Dalai Lama, *The Universe in a Single Atom* (New York: Three Rivers Press, 2008).

3. Blaise Pascal, *Pensées* (New York: Penguin Classics, 1669, 1995).

4. David P. Barash, *Homo Mysterious: Evolutionary Puzzles of Human Nature* (New York: Oxford University Press, 2012).

5. Robert L. Trivers, *The Folly of Fools* (New York: Basic Books, 2011).

6. Bertrand Russell, "A Freeman's Worship," in *The Meaning of Life,* ed. E. D. Klemke and Steven M. Cahn (Oxford: Oxford University Press, 2008). Russell's essay was originally published in 1903.

7. Jean-Paul Sartre, *Existentialism Is a Humanism* (New Haven, CT: Yale University Press, 2007).

8. Ibid.

9. André Malraux, *The Walnut Trees of Altenberg* (New York: Howard Fertig, 1989).

10. Russell, "A Freeman's Worship."

11. Penelope Niven, *Thornton Wilder* (New York: Harper, 2012).

APPENDIX

1. From Thich Nhat Hanh, *Being Peace* (Berkeley, CA: Parallax Press, 1987).

GLOSSARY

AHIMSA—nonkilling, nonhurting, nonviolence (although better envisioned as something positive rather than "non-")

ANATMAN—not-self; denial that things, animals, or people have an independent, substantial nature

ANITYA—impermanence; a state of constantly becoming

ATMAN—the soul or permanent self according to Hinduism

BARDOS—stages between death and one's next incarnation, according to Tibetan Buddhism

BODHISATTVA—one who has achieved substantial enlightenment and seeks to help others do the same

DALAI LAMA—the title of the head of the Geluk School of Tibetan Buddhism

DHARMA—the Buddha's teaching; the law or nature of all things

DOSA—A kind of negative craving, hatred, or aversion

DUKKHA—disappointment; suffering

IDDHI—magical capacities supposedly characteristic of advanced meditators

JATAKAS—stories about the early life (typically in animal form) of the Buddha

KARMA—action; intentional acts and the effect they have on the actor

KUSULA—skillful action, of the sort conducted by enlightened people

MAYA—illusion; error

MAHAYANA—the self-proclaimed "great vehicle" of Buddhism; one of Buddhism's three major divisions, focused especially on enlightenment for all

METTA—loving-kindness

NIRVANA—blissful release from the supposed continuing cycles of birth, death, and rebirth

PRATITYA-SAMUTPADA—the dependent co-arising of all phenomena; dependent origination; interdependence of all things

PRETAS—hungry ghosts, according to Buddhist mythology

RAGA—A particular kind of craving, analogous to desire

SAMATHA—tranquility; calmness; associated with meditative techniques designed to enhance these experiences

SAMSARA—birth, death, and rebirth; the world of suffering

SANGHA—the community of fellow Buddhists

SANKHARAS—mental formations as distinct from fundamental realities

SKANDHAS—the various "aggregates" that together constitute our perceived "selves"

SUKHA—happiness; contentment

SUNYA—hollowness

SUTRA—text; scripture; a collection of the Buddha's discourses

SVABHAVA—unchanging essence that is wrongly thought to constitute living things

TANHA—desires and cravings that lead to *dukkha*

TANTRA—esoteric texts and traditions of Buddhism

THERAVADA—the "lesser vehicle"; a branch of Buddhism that claims to be closest to the Buddha's original teachings and that emphasizes personal enlightenment

TRIVISA—"poisons" that are considered responsible for *dukkha* (q.v.)

VIPASSANA—insight; wisdom; associated with meditative techniques designed to enhance these experiences

pp 37–38: "If you are a poet ... the universe in it." - Reprinted from Heart of Understanding: Commentaries on the Prajnaparamita Heart Sutra (1988) by Thich Nhat Hanh with permission of Parallax Press, Berkeley, California, www.parallax.org.

pp 95–96: "Don't say that I ... the door of compassion." - Reprinted from Call Me By My True Names: The Collected Poems of Thich Nhat Hanh (1999) by Thich Nhat Hanh with permission of Parallax Press, Berkeley, California, www.parallax.org.

pp 187–188: Appendix - Reprinted from Interbeing: Fourteen Guidelines for Engaged Buddhism (1993) by Thich Nhat Hanh with permission of Parallax Press, Berkeley, California, www.parallax.org.

INDEX

"n" following a page number indicates a footnote. Emboldened page ranges refer to a chapter.